(OBD-II Trouble Codes)

All Rights Reserved.

Definitions for generic powertrain diagnostic trouble codes.

(OBD-II Trouble Codes)

P0001	Fuel Volume Regulator Control Circuit/Open
P0002	Fuel Volume Regulator Control Circuit Range/Performance
P0003	Fuel Volume Regulator Control Circuit Low
P0004	Fuel Volume Regulator Control Circuit High
P0005	Fuel Shutoff Valve "A" Control Circuit/Open
P0006	Fuel Shutoff Valve "A" Control Circuit Low
P0007	Fuel Shutoff Valve "A" Control Circuit High
P0008	Engine Position System Performance
P0009	Engine Position System Performance
P0010	"A" Camshaft Position Actuator Circuit
P0011	"A" Camshaft Position - Timing Over-Advanced or System Performance
P0012	"A" Camshaft Position - Timing Over-Retarded
P0013	"B" Camshaft Position - Actuator Circuit
P0014	"B" Camshaft Position - Timing Over-Advanced or System Performance
P0015	"B" Camshaft Position - Timing Over-Retarded
P0016	Crankshaft Position - Camshaft Position Correlation
P0017	Crankshaft Position - Camshaft Position Correlation
P0018	Crankshaft Position - Camshaft Position Correlation
P0019	Crankshaft Position - Camshaft Position Correlation

Vehicle CODE LOG

DATE	
Vehicle Info	
Vehicle CODE	P:
Current problems	
Solution	
NOTE	

P0020	"A" Camshaft Position Actuator Circuit
P0021	"A" Camshaft Position - Timing Over-Advanced or System Performance
P0022	"A" Camshaft Position - Timing Over-Retarded
P0023	"B" Camshaft Position - Actuator Circuit
P0024	"B" Camshaft Position - Timing Over-Advanced or System Performance
P0025	"B" Camshaft Position - Timing Over-Retarded
P0026	Intake Valve Control Solenoid Circuit Range/Performance
P0027	Exhaust Valve Control Solenoid Circuit Range/Performance
P0028	Intake Valve Control Solenoid Circuit Range/Performance
P0029	Exhaust Valve Control Solenoid Circuit Range/Performance
P0030	HO2S Heater Control Circuit
P0031	HO2S Heater Control Circuit Low
P0032	HO2S Heater Control Circuit High
P0033	Turbo Charger Bypass Valve Control Circuit
P0034	Turbo Charger Bypass Valve Control Circuit Low
P0035	Turbo Charger Bypass Valve Control Circuit High
P0036	HO2S Heater Control Circuit
P0037	HO2S Heater Control Circuit Low
P0038	HO2S Heater Control Circuit High
P0039	Turbo/Super Charger Bypass Valve Control Circuit Range/Performance
P0040	O2 Sensor Signals Swapped Bank 1 Sensor 1/ Bank 2 Sensor 1
P0041	O2 Sensor Signals Swapped Bank 1 Sensor 2/ Bank 2 Sensor 2
P0042	HO2S Heater Control Circuit
P0043	HO2S Heater Control Circuit Low
P0044	HO2S Heater Control Circuit High
P0045	Turbo/Super Charger Boost Control Solenoid Circuit/Open
P0046	Turbo/Super Charger Boost Control Solenoid Circuit Range/Performance
P0047	Turbo/Super Charger Boost Control Solenoid Circuit Low
P0048	Turbo/Super Charger Boost Control Solenoid Circuit High
P0049	Turbo/Super Charger Turbine Overspeed
P0050	HO2S Heater Control Circuit
P0051	HO2S Heater Control Circuit Low
P0052	HO2S Heater Control Circuit High

Vehicle CODE LOG

DATE	
Vehicle Info	
Vehicle CODE	P:
Current problems	
Solution	
NOTE	

P0053	HO2S Heater Resistance
P0054	HO2S Heater Resistance
P0055	HO2S Heater Resistance
P0056	HO2S Heater Control Circuit
P0057	HO2S Heater Control Circuit Low
P0058	HO2S Heater Control Circuit High
P0059	HO2S Heater Resistance
P0060	HO2S Heater Resistance
P0061	HO2S Heater Resistance
P0062	HO2S Heater Control Circuit
P0063	HO2S Heater Control Circuit Low
P0064	HO2S Heater Control Circuit High
P0065	Air Assisted Injector Control Range/Performance
P0066	Air Assisted Injector Control Circuit or Circuit Low
P0067	Air Assisted Injector Control Circuit High
P0068	MAP/MAF - Throttle Position Correlation
P0069	Manifold Absolute Pressure - Barometric Pressure Correlation
P0070	Ambient Air Temperature Sensor Circuit
P0071	Ambient Air Temperature Sensor Range/Performance
P0072	Ambient Air Temperature Sensor Circuit Low
P0073	Ambient Air Temperature Sensor Circuit High
P0074	Ambient Air Temperature Sensor Circuit Intermittent
P0075	Intake Valve Control Solenoid Circuit
P0076	Intake Valve Control Solenoid Circuit Low
P0077	Intake Valve Control Solenoid Circuit High
P0078	Exhaust Valve Control Solenoid Circuit
P0079	Exhaust Valve Control Solenoid Circuit Low
P0080	Exhaust Valve Control Solenoid Circuit High
P0081	Intake Valve Control Solenoid Circuit
P0082	Intake Valve Control Solenoid Circuit Low
P0083	Intake Valve Control Solenoid Circuit High
P0084	Exhaust Valve Control Solenoid Circuit
P0085	Exhaust Valve Control Solenoid Circuit Low

Vehicle CODE LOG

DATE	
Vehicle Info	
Vehicle CODE	P:
Current problems	
Solution	
NOTE	

P0086	Exhaust Valve Control Solenoid Circuit High
P0087	Fuel Rail/System Pressure - Too Low
P0088	Fuel Rail/System Pressure - Too High
P0089	Fuel Pressure Regulator 1 Performance
P0090	Fuel Pressure Regulator 1 Control Circuit
P0091	Fuel Pressure Regulator 1 Control Circuit Low
P0092	Fuel Pressure Regulator 1 Control Circuit High
P0093	Fuel System Leak Detected - Large Leak
P0094	Fuel System Leak Detected - Small Leak
P0095	Intake Air Temperature Sensor 2 Circuit
P0096	Intake Air Temperature Sensor 2 Circuit Range/Performance
P0097	Intake Air Temperature Sensor 2 Circuit Low
P0098	Intake Air Temperature Sensor 2 Circuit High
P0099	Intake Air Temperature Sensor 2 Circuit Intermittent/Erratic
P0100	Mass or Volume Air Flow Circuit
P0101	Mass or Volume Air Flow Circuit Range/Performance
P0102	Mass or Volume Air Flow Circuit Low Input
P0103	Mass or Volume Air Flow Circuit High Input
P0104	Mass or Volume Air Flow Circuit Intermittent
P0105	Manifold Absolute Pressure/Barometric Pressure Circuit
P0106	Manifold Absolute Pressure/Barometric Pressure Circuit Range/Performance
P0107	Manifold Absolute Pressure/Barometric Pressure Circuit Low Input
P0108	Manifold Absolute Pressure/Barometric Pressure Circuit High Input
P0109	Manifold Absolute Pressure/Barometric Pressure Circuit Intermittent
P0110	Intake Air Temperature Sensor 1 Circuit
P0111	Intake Air Temperature Sensor 1 Circuit Range/Performance
P0112	Intake Air Temperature Sensor 1 Circuit Low
P0113	Intake Air Temperature Sensor 1 Circuit High
P0114	Intake Air Temperature Sensor 1 Circuit Intermittent
P0115	Engine Coolant Temperature Circuit
P0116	Engine Coolant Temperature Circuit Range/Performance
P0117	Engine Coolant Temperature Circuit Low
P0118	Engine Coolant Temperature Circuit High

Vehicle CODE LOG

DATE	
Vehicle Info	
Vehicle CODE	P:
Current problems	
Solution	
NOTE	

P0119	Engine Coolant Temperature Circuit Intermittent
P0120	Throttle/Pedal Position Sensor/Switch "A" Circuit
P0121	Throttle/Pedal Position Sensor/Switch "A" Circuit Range/Performance
P0122	Throttle/Pedal Position Sensor/Switch "A" Circuit Low
P0123	Throttle/Pedal Position Sensor/Switch "A" Circuit High
P0124	Throttle/Pedal Position Sensor/Switch "A" Circuit Intermittent
P0125	Insufficient Coolant Temperature for Closed Loop Fuel Control
P0126	Insufficient Coolant Temperature for Stable Operation
P0127	Intake Air Temperature Too High
P0128	Coolant Thermostat (Coolant Temperature Below Thermostat Regulating Temperature)
P0129	Barometric Pressure Too Low
P0130	O2 Sensor Circuit
P0131	O2 Sensor Circuit Low Voltage
P0132	O2 Sensor Circuit High Voltage
P0133	O2 Sensor Circuit Slow Response
P0134	O2 Sensor Circuit No Activity Detected
P0135	O2 Sensor Heater Circuit
P0136	O2 Sensor Circuit
P0137	O2 Sensor Circuit Low Voltage
P0138	O2 Sensor Circuit High Voltage
P0139	O2 Sensor Circuit Slow Response
P0140	O2 Sensor Circuit No Activity Detected
P0141	O2 Sensor Heater Circuit
P0142	O2 Sensor Circuit
P0143	O2 Sensor Circuit Low Voltage
P0144	O2 Sensor Circuit High Voltage
P0145	O2 Sensor Circuit Slow Response
P0146	O2 Sensor Circuit No Activity Detected
P0147	O2 Sensor Heater Circuit
P0148	Fuel Delivery Error
P0149	Fuel Timing Error
P0150	O2 Sensor Circuit
P0151	O2 Sensor Circuit Low Voltage

Vehicle CODE LOG

DATE	
Vehicle Info	
Vehicle CODE	P:
Current problems	
Solution	
NOTE	

P0152	O2 Sensor Circuit High Voltage
P0153	O2 Sensor Circuit Slow Response
P0154	O2 Sensor Circuit No Activity Detected
P0155	O2 Sensor Heater Circuit
P0156	O2 Sensor Circuit
P0157	O2 Sensor Circuit Low Voltage
P0158	O2 Sensor Circuit High Voltage
P0159	O2 Sensor Circuit Slow Response
P0160	O2 Sensor Circuit No Activity Detected
P0161	O2 Sensor Heater Circuit
P0162	O2 Sensor Circuit
P0163	O2 Sensor Circuit Low Voltage
P0164	O2 Sensor Circuit High Voltage
P0165	O2 Sensor Circuit Slow Response
P0166	O2 Sensor Circuit No Activity Detected
P0167	O2 Sensor Heater Circuit
P0168	Fuel Temperature Too High
P0169	Incorrect Fuel Composition
P0170	Fuel Trim
P0171	System Too Lean
P0172	System Too Rich
P0173	Fuel Trim
P0174	System Too Lean
P0175	System Too Rich
P0176	Fuel Composition Sensor Circuit
P0177	Fuel Composition Sensor Circuit Range/Performance
P0178	Fuel Composition Sensor Circuit Low
P0179	Fuel Composition Sensor Circuit High
P0180	Fuel Temperature Sensor A Circuit
P0181	Fuel Temperature Sensor A Circuit Range/Performance
P0182	Fuel Temperature Sensor A Circuit Low
P0183	Fuel Temperature Sensor A Circuit High
P0184	Fuel Temperature Sensor A Circuit Intermittent

Vehicle CODE LOG

DATE	
Vehicle Info	
Vehicle CODE	P:
Current problems	
Solution	
NOTE	

P0185	Fuel Temperature Sensor B Circuit
P0186	Fuel Temperature Sensor B Circuit Range/Performance
P0187	Fuel Temperature Sensor B Circuit Low
P0188	Fuel Temperature Sensor B Circuit High
P0189	Fuel Temperature Sensor B Circuit Intermittent
P0190	Fuel Rail Pressure Sensor Circuit
P0191	Fuel Rail Pressure Sensor Circuit Range/Performance
P0192	Fuel Rail Pressure Sensor Circuit Low
P0193	Fuel Rail Pressure Sensor Circuit High
P0194	Fuel Rail Pressure Sensor Circuit Intermittent
P0195	Engine Oil Temperature Sensor
P0196	Engine Oil Temperature Sensor Range/Performance
P0197	Engine Oil Temperature Sensor Low
P0198	Engine Oil Temperature Sensor High
P0199	Engine Oil Temperature Sensor Intermittent
P0200	Injector Circuit/Open
P0201	Injector Circuit/Open - Cylinder 1
P0202	Injector Circuit/Open - Cylinder 2
P0203	Injector Circuit/Open - Cylinder 3
P0204	Injector Circuit/Open - Cylinder 4
P0205	Injector Circuit/Open - Cylinder 5
P0206	Injector Circuit/Open - Cylinder 6
P0207	Injector Circuit/Open - Cylinder 7
P0208	Injector Circuit/Open - Cylinder 8
P0209	Injector Circuit/Open - Cylinder 9
P0210	Injector Circuit/Open - Cylinder 10
P0211	Injector Circuit/Open - Cylinder 11
P0212	Injector Circuit/Open - Cylinder 12
P0213	Cold Start Injector 1
P0214	Cold Start Injector 2
P0215	Engine Shutoff Solenoid
P0216	Injector/Injection Timing Control Circuit
P0217	Engine Coolant Over Temperature Condition

Vehicle CODE LOG

DATE	
Vehicle Info	
Vehicle CODE	P:
Current problems	
Solution	
NOTE	

P0218	Transmission Fluid Over Temperature Condition
P0219	Engine Overspeed Condition
P0220	Throttle/Pedal Position Sensor/Switch "B" Circuit
P0221	Throttle/Pedal Position Sensor/Switch "B" Circuit Range/Performance
P0222	Throttle/Pedal Position Sensor/Switch "B" Circuit Low
P0223	Throttle/Pedal Position Sensor/Switch "B" Circuit High
P0224	Throttle/Pedal Position Sensor/Switch "B" Circuit Intermittent
P0225	Throttle/Pedal Position Sensor/Switch "C" Circuit
P0226	Throttle/Pedal Position Sensor/Switch "C" Circuit Range/Performance
P0227	Throttle/Pedal Position Sensor/Switch "C" Circuit Low
P0228	Throttle/Pedal Position Sensor/Switch "C" Circuit High
P0229	Throttle/Pedal Position Sensor/Switch "C" Circuit Intermittent
P0230	Fuel Pump Primary Circuit
P0231	Fuel Pump Secondary Circuit Low
P0232	Fuel Pump Secondary Circuit High
P0233	Fuel Pump Secondary Circuit Intermittent
P0234	Turbo/Super Charger Overboost Condition
P0235	Turbo/Super Charger Boost Sensor "A" Circuit
P0236	Turbo/Super Charger Boost Sensor "A" Circuit Range/Performance
P0237	Turbo/Super Charger Boost Sensor "A" Circuit Low
P0238	Turbo/Super Charger Boost Sensor "A" Circuit High
P0239	Turbo/Super Charger Boost Sensor "B" Circuit
P0240	Turbo/Super Charger Boost Sensor "B" Circuit Range/Performance
P0241	Turbo/Super Charger Boost Sensor "B" Circuit Low
P0242	Turbo/Super Charger Boost Sensor "B" Circuit High
P0243	Turbo/Super Charger Wastegate Solenoid "A"
P0244	Turbo/Super Charger Wastegate Solenoid "A" Range/Performance
P0245	Turbo/Super Charger Wastegate Solenoid "A" Low
P0246	Turbo/Super Charger Wastegate Solenoid "A" High
P0247	Turbo/Super Charger Wastegate Solenoid "B"
P0248	Turbo/Super Charger Wastegate Solenoid "B" Range/Performance
P0249	Turbo/Super Charger Wastegate Solenoid "B" Low
P0250	Turbo/Super Charger Wastegate Solenoid "B" High

Vehicle CODE LOG

DATE	
Vehicle Info	
Vehicle CODE	P:
Current problems	
Solution	
NOTE	

P0251	Injection Pump Fuel Metering Control "A" (Cam/Rotor/Injector)
P0252	Injection Pump Fuel Metering Control "A" Range/Performance (Cam/Rotor/Injector)
P0253	Injection Pump Fuel Metering Control "A" Low (Cam/Rotor/Injector)
P0254	Injection Pump Fuel Metering Control "A" High (Cam/Rotor/Injector)
P0255	Injection Pump Fuel Metering Control "A" Intermittent (Cam/Rotor/Injector)
P0256	Injection Pump Fuel Metering Control "B" (Cam/Rotor/Injector)
P0257	Injection Pump Fuel Metering Control "B" Range/Performance (Cam/Rotor/Injector)
P0258	Injection Pump Fuel Metering Control "B" Low (Cam/Rotor/Injector)
P0259	Injection Pump Fuel Metering Control "B" High (Cam/Rotor/Injector)
P0260	Injection Pump Fuel Metering Control "B" Intermittent (Cam/Rotor/Injector)
P0261	Cylinder 1 Injector Circuit Low
P0262	Cylinder 1 Injector Circuit High
P0263	Cylinder 1 Contribution/Balance
P0264	Cylinder 2 Injector Circuit Low
P0265	Cylinder 2 Injector Circuit High
P0266	Cylinder 2 Contribution/Balance
P0267	Cylinder 3 Injector Circuit Low
P0268	Cylinder 3 Injector Circuit High
P0269	Cylinder 3 Contribution/Balance
P0270	Cylinder 4 Injector Circuit Low
P0271	Cylinder 4 Injector Circuit High
P0272	Cylinder 4 Contribution/Balance
P0273	Cylinder 5 Injector Circuit Low
P0274	Cylinder 5 Injector Circuit High
P0275	Cylinder 5 Contribution/Balance
P0276	Cylinder 6 Injector Circuit Low
P0277	Cylinder 6 Injector Circuit High
P0278	Cylinder 6 Contribution/Balance
P0279	Cylinder 7 Injector Circuit Low
P0280	Cylinder 7 Injector Circuit High
P0281	Cylinder 7 Contribution/Balance
P0282	Cylinder 8 Injector Circuit Low
P0283	Cylinder 8 Injector Circuit High

Vehicle CODE LOG

DATE	
Vehicle Info	
Vehicle CODE	P:
Current problems	
Solution	
NOTE	

P0284	Cylinder 8 Contribution/Balance
P0285	Cylinder 9 Injector Circuit Low
P0286	Cylinder 9 Injector Circuit High
P0287	Cylinder 9 Contribution/Balance
P0288	Cylinder 10 Injector Circuit Low
P0289	Cylinder 10 Injector Circuit High
P0290	Cylinder 10 Contribution/Balance
P0291	Cylinder 11 Injector Circuit Low
P0292	Cylinder 11 Injector Circuit High
P0293	Cylinder 11 Contribution/Balance
P0294	Cylinder 12 Injector Circuit Low
P0295	Cylinder 12 Injector Circuit High
P0296	Cylinder 12 Contribution/Balance
P0297	Vehicle Overspeed Condition
P0298	Engine Oil Over Temperature
P0299	Turbo/Super Charger Underboost
P0300	Random/Multiple Cylinder Misfire Detected
P0301	Cylinder 1 Misfire Detected
P0302	Cylinder 2 Misfire Detected
P0303	Cylinder 3 Misfire Detected
P0304	Cylinder 4 Misfire Detected
P0305	Cylinder 5 Misfire Detected
P0306	Cylinder 6 Misfire Detected
P0307	Cylinder 7 Misfire Detected
P0308	Cylinder 8 Misfire Detected
P0309	Cylinder 9 Misfire Detected
P0310	Cylinder 10 Misfire Detected
P0311	Cylinder 11 Misfire Detected
P0312	Cylinder 12 Misfire Detected
P0313	Misfire Detected with Low Fuel
P0314	Single Cylinder Misfire (Cylinder not Specified)
P0315	Crankshaft Position System Variation Not Learned
P0316	Engine Misfire Detected on Startup (First 1000 Revolutions)

Vehicle CODE LOG

DATE	
Vehicle Info	
Vehicle CODE	P:
Current problems	
Solution	
NOTE	

P0317	Rough Road Hardware Not Present
P0318	Rough Road Sensor "A" Signal Circuit
P0319	Rough Road Sensor "B"
P0320	Ignition/Distributor Engine Speed Input Circuit
P0321	Ignition/Distributor Engine Speed Input Circuit Range/Performance
P0322	Ignition/Distributor Engine Speed Input Circuit No Signal
P0323	Ignition/Distributor Engine Speed Input Circuit Intermittent
P0324	Knock Control System Error
P0325	Knock Sensor 1 Circuit
P0326	Knock Sensor 1 Circuit Range/Performance
P0327	Knock Sensor 1 Circuit Low
P0328	Knock Sensor 1 Circuit High
P0329	Knock Sensor 1 Circuit Input Intermittent
P0330	Knock Sensor 2 Circuit
P0331	Knock Sensor 2 Circuit Range/Performance
P0332	Knock Sensor 2 Circuit Low
P0333	Knock Sensor 2 Circuit High
P0334	Knock Sensor 2 Circuit Input Intermittent
P0335	Crankshaft Position Sensor "A" Circuit
P0336	Crankshaft Position Sensor "A" Circuit Range/Performance
P0337	Crankshaft Position Sensor "A" Circuit Low
P0338	Crankshaft Position Sensor "A" Circuit High
P0339	Crankshaft Position Sensor "A" Circuit Intermittent
P0340	Camshaft Position Sensor "A" Circuit
P0341	Camshaft Position Sensor "A" Circuit Range/Performance
P0342	Camshaft Position Sensor "A" Circuit Low
P0343	Camshaft Position Sensor "A" Circuit High
P0344	Camshaft Position Sensor "A" Circuit Intermittent
P0345	Camshaft Position Sensor "A" Circuit
P0346	Camshaft Position Sensor "A" Circuit Range/Performance
P0347	Camshaft Position Sensor "A" Circuit Low
P0348	Camshaft Position Sensor "A" Circuit High
P0349	Camshaft Position Sensor "A" Circuit Intermittent

Vehicle CODE LOG

DATE	
Vehicle Info	
Vehicle CODE	P:
Current problems	
Solution	
NOTE	

P0350	Ignition Coil Primary/Secondary Circuit
P0351	Ignition Coil "A" Primary/Secondary Circuit
P0352	Ignition Coil "B" Primary/Secondary Circuit
P0353	Ignition Coil "C" Primary/Secondary Circuit
P0354	Ignition Coil "D" Primary/Secondary Circuit
P0355	Ignition Coil "E" Primary/Secondary Circuit
P0356	Ignition Coil "F" Primary/Secondary Circuit
P0357	Ignition Coil "G" Primary/Secondary Circuit
P0358	Ignition Coil "H" Primary/Secondary Circuit
P0359	Ignition Coil "I" Primary/Secondary Circuit
P0360	Ignition Coil "J" Primary/Secondary Circuit
P0361	Ignition Coil "K" Primary/Secondary Circuit
P0362	Ignition Coil "L" Primary/Secondary Circuit
P0363	Misfire Detected - Fueling Disabled
P0364	Reserved
P0365	Camshaft Position Sensor "B" Circuit
P0366	Camshaft Position Sensor "B" Circuit Range/Performance
P0367	Camshaft Position Sensor "B" Circuit Low
P0368	Camshaft Position Sensor "B" Circuit High
P0369	Camshaft Position Sensor "B" Circuit Intermittent
P0370	Timing Reference High Resolution Signal "A"
P0371	Timing Reference High Resolution Signal "A" Too Many Pulses
P0372	Timing Reference High Resolution Signal "A" Too Few Pulses
P0373	Timing Reference High Resolution Signal "A" Intermittent/Erratic Pulses
P0374	Timing Reference High Resolution Signal "A" No Pulse
P0375	Timing Reference High Resolution Signal "B"
P0376	Timing Reference High Resolution Signal "B" Too Many Pulses
P0377	Timing Reference High Resolution Signal "B" Too Few Pulses
P0378	Timing Reference High Resolution Signal "B" Intermittent/Erratic Pulses
P0379	Timing Reference High Resolution Signal "B" No Pulses
P0380	Glow Plug/Heater Circuit "A"
P0381	Glow Plug/Heater Indicator Circuit
P0382	Glow Plug/Heater Circuit "B"

Vehicle CODE LOG

DATE	
Vehicle Info	
Vehicle CODE	P:
Current problems	
Solution	
NOTE	

P0383	Reserved by SAE J2012
P0384	Reserved by SAE J2012
P0385	Crankshaft Position Sensor "B" Circuit
P0386	Crankshaft Position Sensor "B" Circuit Range/Performance
P0387	Crankshaft Position Sensor "B" Circuit Low
P0388	Crankshaft Position Sensor "B" Circuit High
P0389	Crankshaft Position Sensor "B" Circuit Intermittent
P0390	Camshaft Position Sensor "B" Circuit
P0391	Camshaft Position Sensor "B" Circuit Range/Performance
P0392	Camshaft Position Sensor "B" Circuit Low
P0393	Camshaft Position Sensor "B" Circuit High
P0394	Camshaft Position Sensor "B" Circuit Intermittent
P0400	Exhaust Gas Recirculation Flow
P0401	Exhaust Gas Recirculation Flow Insufficient Detected
P0402	Exhaust Gas Recirculation Flow Excessive Detected
P0403	Exhaust Gas Recirculation Control Circuit
P0404	Exhaust Gas Recirculation Control Circuit Range/Performance
P0405	Exhaust Gas Recirculation Sensor "A" Circuit Low
P0406	Exhaust Gas Recirculation Sensor "A" Circuit High
P0407	Exhaust Gas Recirculation Sensor "B" Circuit Low
P0408	Exhaust Gas Recirculation Sensor "B" Circuit High
P0409	Exhaust Gas Recirculation Sensor "A" Circuit
P0410	Secondary Air Injection System
P0411	Secondary Air Injection System Incorrect Flow Detected
P0412	Secondary Air Injection System Switching Valve "A" Circuit
P0413	Secondary Air Injection System Switching Valve "A" Circuit Open
P0414	Secondary Air Injection System Switching Valve "A" Circuit Shorted
P0415	Secondary Air Injection System Switching Valve "B" Circuit
P0416	Secondary Air Injection System Switching Valve "B" Circuit Open
P0417	Secondary Air Injection System Switching Valve "B" Circuit Shorted
P0418	Secondary Air Injection System Control "A" Circuit
P0419	Secondary Air Injection System Control "B" Circuit
P0420	Catalyst System Efficiency Below Threshold

Vehicle CODE LOG

DATE	
Vehicle Info	
Vehicle CODE	P:
Current problems	
Solution	
NOTE	

P0421	Warm Up Catalyst Efficiency Below Threshold
P0422	Main Catalyst Efficiency Below Threshold
P0423	Heated Catalyst Efficiency Below Threshold
P0424	Heated Catalyst Temperature Below Threshold
P0425	Catalyst Temperature Sensor
P0426	Catalyst Temperature Sensor Range/Performance
P0427	Catalyst Temperature Sensor Low
P0428	Catalyst Temperature Sensor High
P0429	Catalyst Heater Control Circuit
P0430	Catalyst System Efficiency Below Threshold
P0431	Warm Up Catalyst Efficiency Below Threshold
P0432	Main Catalyst Efficiency Below Threshold
P0433	Heated Catalyst Efficiency Below Threshold
P0434	Heated Catalyst Temperature Below Threshold
P0435	Catalyst Temperature Sensor
P0436	Catalyst Temperature Sensor Range/Performance
P0437	Catalyst Temperature Sensor Low
P0438	Catalyst Temperature Sensor High
P0439	Catalyst Heater Control Circuit
P0440	Evaporative Emission System
P0441	Evaporative Emission System Incorrect Purge Flow
P0442	Evaporative Emission System Leak Detected (small leak)
P0443	Evaporative Emission System Purge Control Valve Circuit
P0444	Evaporative Emission System Purge Control Valve Circuit Open
P0445	Evaporative Emission System Purge Control Valve Circuit Shorted
P0446	Evaporative Emission System Vent Control Circuit
P0447	Evaporative Emission System Vent Control Circuit Open
P0448	Evaporative Emission System Vent Control Circuit Shorted
P0449	Evaporative Emission System Vent Valve/Solenoid Circuit
P0450	Evaporative Emission System Pressure Sensor/Switch
P0451	Evaporative Emission System Pressure Sensor/Switch Range/Performance
P0452	Evaporative Emission System Pressure Sensor/Switch Low
P0453	Evaporative Emission System Pressure Sensor/Switch High

Vehicle CODE LOG

DATE	
Vehicle Info	
Vehicle CODE	P:
Current problems	
Solution	
NOTE	

P0454	Evaporative Emission System Pressure Sensor/Switch Intermittent
P0455	Evaporative Emission System Leak Detected (large leak)
P0456	Evaporative Emission System Leak Detected (very small leak)
P0457	Evaporative Emission System Leak Detected (fuel cap loose/off)
P0458	Evaporative Emission System Purge Control Valve Circuit Low
P0459	Evaporative Emission System Purge Control Valve Circuit High
P0460	Fuel Level Sensor "A" Circuit
P0461	Fuel Level Sensor "A" Circuit Range/Performance
P0462	Fuel Level Sensor "A" Circuit Low
P0463	Fuel Level Sensor "A" Circuit High
P0464	Fuel Level Sensor "A" Circuit Intermittent
P0465	EVAP Purge Flow Sensor Circuit
P0466	EVAP Purge Flow Sensor Circuit Range/Performance
P0467	EVAP Purge Flow Sensor Circuit Low
P0468	EVAP Purge Flow Sensor Circuit High
P0469	EVAP Purge Flow Sensor Circuit Intermittent
P0470	Exhaust Pressure Sensor
P0471	Exhaust Pressure Sensor Range/Performance
P0472	Exhaust Pressure Sensor Low
P0473	Exhaust Pressure Sensor High
P0474	Exhaust Pressure Sensor Intermittent
P0475	Exhaust Pressure Control Valve
P0476	Exhaust Pressure Control Valve Range/Performance
P0477	Exhaust Pressure Control Valve Low
P0478	Exhaust Pressure Control Valve High
P0479	Exhaust Pressure Control Valve Intermittent
P0480	Fan 1 Control Circuit
P0481	Fan 2 Control Circuit
P0482	Fan 3 Control Circuit
P0483	Fan Rationality Check
P0484	Fan Circuit Over Current
P0485	Fan Power/Ground Circuit
P0486	Exhaust Gas Recirculation Sensor "B" Circuit

Vehicle CODE LOG

DATE	
Vehicle Info	
Vehicle CODE	P:
Current problems	
Solution	
NOTE	

P0487	Exhaust Gas Recirculation Throttle Position Control Circuit
P0488	Exhaust Gas Recirculation Throttle Position Control Range/Performance
P0489	Exhaust Gas Recirculation Control Circuit Low
P0490	Exhaust Gas Recirculation Control Circuit High
P0491	Secondary Air Injection System Insufficient Flow
P0492	Secondary Air Injection System Insufficient Flow
P0493	Fan Overspeed
P0494	Fan Speed Low
P0495	Fan Speed High
P0496	Evaporative Emission System High Purge Flow
P0497	Evaporative Emission System Low Purge Flow
P0498	Evaporative Emission System Vent Valve Control Circuit Low
P0499	Evaporative Emission System Vent Valve Control Circuit High
P0500	Vehicle Speed Sensor "A"
P0501	Vehicle Speed Sensor "A" Range/Performance
P0502	Vehicle Speed Sensor "A" Circuit Low Input
P0503	Vehicle Speed Sensor "A" Intermittent/Erratic/High
P0504	Brake Switch "A"/"B" Correlation
P0505	Idle Air Control System
P0506	Idle Air Control System RPM Lower Than Expected
P0507	Idle Air Control System RPM Higher Than Expected
P0508	Idle Air Control System Circuit Low
P0509	Idle Air Control System Circuit High
P0510	Closed Throttle Position Switch
P0511	Idle Air Control Circuit
P0512	Starter Request Circuit
P0513	Incorrect Immobilizer Key
P0514	Battery Temperature Sensor Circuit Range/Performance
P0515	Battery Temperature Sensor Circuit
P0516	Battery Temperature Sensor Circuit Low
P0517	Battery Temperature Sensor Circuit High
P0518	Idle Air Control Circuit Intermittent
P0519	Idle Air Control System Performance

Vehicle CODE LOG

DATE	
Vehicle Info	
Vehicle CODE	P:
Current problems	
Solution	
NOTE	

P0520	Engine Oil Pressure Sensor/Switch Circuit
P0521	Engine Oil Pressure Sensor/Switch Range/Performance
P0522	Engine Oil Pressure Sensor/Switch Low Voltage
P0523	Engine Oil Pressure Sensor/Switch High Voltage
P0524	Engine Oil Pressure Too Low
P0525	Cruise Control Servo Control Circuit Range/Performance
P0526	Fan Speed Sensor Circuit
P0527	Fan Speed Sensor Circuit Range/Performance
P0528	Fan Speed Sensor Circuit No Signal
P0529	Fan Speed Sensor Circuit Intermittent
P0530	A/C Refrigerant Pressure Sensor "A" Circuit
P0531	A/C Refrigerant Pressure Sensor "A" Circuit Range/Performance
P0532	A/C Refrigerant Pressure Sensor "A" Circuit Low
P0533	A/C Refrigerant Pressure Sensor "A" Circuit High
P0534	Air Conditioner Refrigerant Charge Loss
P0535	A/C Evaporator Temperature Sensor Circuit
P0536	A/C Evaporator Temperature Sensor Circuit Range/Performance
P0537	A/C Evaporator Temperature Sensor Circuit Low
P0538	A/C Evaporator Temperature Sensor Circuit High
P0539	A/C Evaporator Temperature Sensor Circuit Intermittent
P0540	Intake Air Heater "A" Circuit
P0541	Intake Air Heater "A" Circuit Low
P0542	Intake Air Heater "A" Circuit High
P0543	Intake Air Heater "A" Circuit Open
P0544	Exhaust Gas Temperature Sensor Circuit
P0545	Exhaust Gas Temperature Sensor Circuit Low
P0546	Exhaust Gas Temperature Sensor Circuit High
P0547	Exhaust Gas Temperature Sensor Circuit
P0548	Exhaust Gas Temperature Sensor Circuit Low
P0549	Exhaust Gas Temperature Sensor Circuit High
P0550	Power Steering Pressure Sensor/Switch Circuit
P0551	Power Steering Pressure Sensor/Switch Circuit Range/Performance
P0552	Power Steering Pressure Sensor/Switch Circuit Low Input

Vehicle CODE LOG

DATE	
Vehicle Info	
Vehicle CODE	P:
Current problems	
Solution	
NOTE	

P0553	Power Steering Pressure Sensor/Switch Circuit High Input
P0554	Power Steering Pressure Sensor/Switch Circuit Intermittent
P0555	Brake Booster Pressure Sensor Circuit
P0556	Brake Booster Pressure Sensor Circuit Range/Performance
P0557	Brake Booster Pressure Sensor Circuit Low Input
P0558	Brake Booster Pressure Sensor Circuit High Input
P0559	Brake Booster Pressure Sensor Circuit Intermittent
P0560	System Voltage
P0561	System Voltage Unstable
P0562	System Voltage Low
P0563	System Voltage High
P0564	Cruise Control Multi-Function Input "A" Circuit
P0565	Cruise Control On Signal
P0566	Cruise Control Off Signal
P0567	Cruise Control Resume Signal
P0568	Cruise Control Set Signal
P0569	Cruise Control Coast Signal
P0570	Cruise Control Accelerate Signal
P0571	Brake Switch "A" Circuit
P0572	Brake Switch "A" Circuit Low
P0573	Brake Switch "A" Circuit High
P0574	Cruise Control System - Vehicle Speed Too High
P0575	Cruise Control Input Circuit
P0576	Cruise Control Input Circuit Low
P0577	Cruise Control Input Circuit High
P0578	Cruise Control Multi-Function Input "A" Circuit Stuck
P0579	Cruise Control Multi-Function Input "A" Circuit Range/Performance
P0580	Cruise Control Multi-Function Input "A" Circuit Low
P0581	Cruise Control Multi-Function Input "A" Circuit High
P0582	Cruise Control Vacuum Control Circuit/Open
P0583	Cruise Control Vacuum Control Circuit Low
P0584	Cruise Control Vacuum Control Circuit High
P0585	Cruise Control Multi-Function Input "A"/"B" Correlation

Vehicle CODE LOG

DATE	
Vehicle Info	
Vehicle CODE	P:
Current problems	
Solution	
NOTE	

P0586	Cruise Control Vent Control Circuit/Open
P0587	Cruise Control Vent Control Circuit Low
P0588	Cruise Control Vent Control Circuit High
P0589	Cruise Control Multi-Function Input "B" Circuit
P0590	Cruise Control Multi-Function Input "B" Circuit Stuck
P0591	Cruise Control Multi-Function Input "B" Circuit Range/Performance
P0592	Cruise Control Multi-Function Input "B" Circuit Low
P0593	Cruise Control Multi-Function Input "B" Circuit High
P0594	Cruise Control Servo Control Circuit/Open
P0595	Cruise Control Servo Control Circuit Low
P0596	Cruise Control Servo Control Circuit High
P0597	Thermostat Heater Control Circuit/Open
P0598	Thermostat Heater Control Circuit Low
P0599	Thermostat Heater Control Circuit High
P0600	Serial Communication Link
P0601	Internal Control Module Memory Check Sum Error
P0602	Control Module Programming Error
P0603	Internal Control Module Keep Alive Memory (KAM) Error
P0604	Internal Control Module Random Access Memory (RAM) Error
P0605	Internal Control Module Read Only Memory (ROM) Error
P0606	ECM/PCM Processor
P0607	Control Module Performance
P0608	Control Module VSS Output "A"
P0609	Control Module VSS Output "B"
P0610	Control Module Vehicle Options Error
P0611	Fuel Injector Control Module Performance
P0612	Fuel Injector Control Module Relay Control
P0613	TCM Processor
P0614	ECM / TCM Incompatible
P0615	Starter Relay Circuit
P0616	Starter Relay Circuit Low
P0617	Starter Relay Circuit High
P0618	Alternative Fuel Control Module KAM Error

Vehicle CODE LOG

DATE	
Vehicle Info	
Vehicle CODE	P:
Current problems	
Solution	
NOTE	

P0619	Alternative Fuel Control Module RAM/ROM Error
P0620	Generator Control Circuit
P0621	Generator Lamp/L Terminal Circuit
P0622	Generator Field/F Terminal Circuit
P0623	Generator Lamp Control Circuit
P0624	Fuel Cap Lamp Control Circuit
P0625	Generator Field/F Terminal Circuit Low
P0626	Generator Field/F Terminal Circuit High
P0627	Fuel Pump "A" Control Circuit /Open
P0628	Fuel Pump "A" Control Circuit Low
P0629	Fuel Pump "A" Control Circuit High
P0630	VIN Not Programmed or Incompatible - ECM/PCM
P0631	VIN Not Programmed or Incompatible - TCM
P0632	Odometer Not Programmed - ECM/PCM
P0633	Immobilizer Key Not Programmed - ECM/PCM
P0634	PCM/ECM/TCM Internal Temperature Too High
P0635	Power Steering Control Circuit
P0636	Power Steering Control Circuit Low
P0637	Power Steering Control Circuit High
P0638	Throttle Actuator Control Range/Performance
P0639	Throttle Actuator Control Range/Performance
P0640	Intake Air Heater Control Circuit
P0641	Sensor Reference Voltage "A" Circuit/Open
P0642	Sensor Reference Voltage "A" Circuit Low
P0643	Sensor Reference Voltage "A" Circuit High
P0644	Driver Display Serial Communication Circuit
P0645	A/C Clutch Relay Control Circuit
P0646	A/C Clutch Relay Control Circuit Low
P0647	A/C Clutch Relay Control Circuit High
P0648	Immobilizer Lamp Control Circuit
P0649	Speed Control Lamp Control Circuit
P0650	Malfunction Indicator Lamp (MIL) Control Circuit
P0651	Sensor Reference Voltage "B" Circuit/Open

Vehicle CODE LOG

DATE	
Vehicle Info	
Vehicle CODE	P:
Current problems	
Solution	
NOTE	

P0652	Sensor Reference Voltage "B" Circuit Low
P0653	Sensor Reference Voltage "B" Circuit High
P0654	Engine RPM Output Circuit
P0655	Engine Hot Lamp Output Control Circuit
P0656	Fuel Level Output Circuit
P0657	Actuator Supply Voltage "A" Circuit/Open
P0658	Actuator Supply Voltage "A" Circuit Low
P0659	Actuator Supply Voltage "A" Circuit High
P0660	Intake Manifold Tuning Valve Control Circuit/Open
P0661	Intake Manifold Tuning Valve Control Circuit Low
P0662	Intake Manifold Tuning Valve Control Circuit High
P0663	Intake Manifold Tuning Valve Control Circuit/Open
P0664	Intake Manifold Tuning Valve Control Circuit Low
P0665	Intake Manifold Tuning Valve Control Circuit High
P0666	PCM/ECM/TCM Internal Temperature Sensor Circuit
P0667	PCM/ECM/TCM Internal Temperature Sensor Range/Performance
P0668	PCM/ECM/TCM Internal Temperature Sensor Circuit Low
P0669	PCM/ECM/TCM Internal Temperature Sensor Circuit High
P0670	Glow Plug Module Control Circuit
P0671	Cylinder 1 Glow Plug Circuit
P0672	Cylinder 2 Glow Plug Circuit
P0673	Cylinder 3 Glow Plug Circuit
P0674	Cylinder 4 Glow Plug Circuit
P0675	Cylinder 5 Glow Plug Circuit
P0676	Cylinder 6 Glow Plug Circuit
P0677	Cylinder 7 Glow Plug Circuit
P0678	Cylinder 8 Glow Plug Circuit
P0679	Cylinder 9 Glow Plug Circuit
P0680	Cylinder 10 Glow Plug Circuit
P0681	Cylinder 11 Glow Plug Circuit
P0682	Cylinder 12 Glow Plug Circuit
P0683	Glow Plug Control Module to PCM Communication Circuit
P0684	Glow Plug Control Module to PCM Communication Circuit Range/Performance

Vehicle CODE LOG

DATE	
Vehicle Info	
Vehicle CODE	P:
Current problems	
Solution	
NOTE	

P0685	ECM/PCM Power Relay Control Circuit /Open
P0686	ECM/PCM Power Relay Control Circuit Low
P0687	ECM/PCM Power Relay Control Circuit High
P0688	ECM/PCM Power Relay Sense Circuit /Open
P0689	ECM/PCM Power Relay Sense Circuit Low
P0690	ECM/PCM Power Relay Sense Circuit High
P0691	Fan 1 Control Circuit Low
P0692	Fan 1 Control Circuit High
P0693	Fan 2 Control Circuit Low
P0694	Fan 2 Control Circuit High
P0695	Fan 3 Control Circuit Low
P0696	Fan 3 Control Circuit High
P0697	Sensor Reference Voltage "C" Circuit/Open
P0698	Sensor Reference Voltage "C" Circuit Low
P0699	Sensor Reference Voltage "C" Circuit High
P0700	Transmission Control System (MIL Request)
P0701	Transmission Control System Range/Performance
P0702	Transmission Control System Electrical
P0703	Brake Switch "B" Circuit
P0704	Clutch Switch Input Circuit Malfunction
P0705	Transmission Range Sensor Circuit Malfunction (PRNDL Input)
P0706	Transmission Range Sensor Circuit Range/Performance
P0707	Transmission Range Sensor Circuit Low
P0708	Transmission Range Sensor Circuit High
P0709	Transmission Range Sensor Circuit Intermittent
P0710	Transmission Fluid Temperature Sensor "A" Circuit
P0711	Transmission Fluid Temperature Sensor "A" Circuit Range/Performance
P0712	Transmission Fluid Temperature Sensor "A" Circuit Low
P0713	Transmission Fluid Temperature Sensor "A" Circuit High
P0714	Transmission Fluid Temperature Sensor "A" Circuit Intermittent
P0715	Input/Turbine Speed Sensor "A" Circuit
P0716	Input/Turbine Speed Sensor "A" Circuit Range/Performance
P0717	Input/Turbine Speed Sensor "A" Circuit No Signal

Vehicle CODE LOG

DATE	
Vehicle Info	
Vehicle CODE	P:
Current problems	
Solution	
NOTE	

P0718	Input/Turbine Speed Sensor "A" Circuit Intermittent
P0719	Brake Switch "B" Circuit Low
P0720	Output Speed Sensor Circuit
P0721	Output Speed Sensor Circuit Range/Performance
P0722	Output Speed Sensor Circuit No Signal
P0723	Output Speed Sensor Circuit Intermittent
P0724	Brake Switch "B" Circuit High
P0725	Engine Speed Input Circuit
P0726	Engine Speed Input Circuit Range/Performance
P0727	Engine Speed Input Circuit No Signal
P0728	Engine Speed Input Circuit Intermittent
P0729	Gear 6 Incorrect Ratio
P0730	Incorrect Gear Ratio
P0731	Gear 1 Incorrect Ratio
P0732	Gear 2 Incorrect Ratio
P0733	Gear 3 Incorrect Ratio
P0734	Gear 4 Incorrect Ratio
P0735	Gear 5 Incorrect Ratio
P0736	Reverse Incorrect Ratio
P0737	TCM Engine Speed Output Circuit
P0738	TCM Engine Speed Output Circuit Low
P0739	TCM Engine Speed Output Circuit High
P0740	Torque Converter Clutch Circuit/Open
P0741	Torque Converter Clutch Circuit Performance or Stuck Off
P0742	Torque Converter Clutch Circuit Stuck On
P0743	Torque Converter Clutch Circuit Electrical
P0744	Torque Converter Clutch Circuit Intermittent
P0745	Pressure Control Solenoid "A"
P0746	Pressure Control Solenoid "A" Performance or Stuck Off
P0747	Pressure Control Solenoid "A" Stuck On
P0748	Pressure Control Solenoid "A" Electrical
P0749	Pressure Control Solenoid "A" Intermittent
P0750	Shift Solenoid "A"

Vehicle CODE LOG

DATE	
Vehicle Info	
Vehicle CODE	P:
Current problems	
Solution	
NOTE	

P0751	Shift Solenoid "A" Performance or Stuck Off
P0752	Shift Solenoid "A" Stuck On
P0753	Shift Solenoid "A" Electrical
P0754	Shift Solenoid "A" Intermittent
P0755	Shift Solenoid "B"
P0756	Shift Solenoid "B" Performance or Stuck Off
P0757	Shift Solenoid "B" Stuck On
P0758	Shift Solenoid "B" Electrical
P0759	Shift Solenoid "B" Intermittent
P0760	Shift Solenoid "C"
P0761	Shift Solenoid "C" Performance or Stuck Off
P0762	Shift Solenoid "C" Stuck On
P0763	Shift Solenoid "C" Electrical
P0764	Shift Solenoid "C" Intermittent
P0765	Shift Solenoid "D"
P0766	Shift Solenoid "D" Performance or Stuck Off
P0767	Shift Solenoid "D" Stuck On
P0768	Shift Solenoid "D" Electrical
P0769	Shift Solenoid "D" Intermittent
P0770	Shift Solenoid "E"
P0771	Shift Solenoid "E" Performance or Stuck Off
P0772	Shift Solenoid "E" Stuck On
P0773	Shift Solenoid "E" Electrical
P0774	Shift Solenoid "E" Intermittent
P0775	Pressure Control Solenoid "B"
P0776	Pressure Control Solenoid "B" Performance or Stuck off
P0777	Pressure Control Solenoid "B" Stuck On
P0778	Pressure Control Solenoid "B" Electrical
P0779	Pressure Control Solenoid "B" Intermittent
P0780	Shift Error
P0781	1-2 Shift
P0782	2-3 Shift
P0783	3-4 Shift

Vehicle CODE LOG

DATE	
Vehicle Info	
Vehicle CODE	P:
Current problems	
Solution	
NOTE	

P0784	4-5 Shift
P0785	Shift/Timing Solenoid
P0786	Shift/Timing Solenoid Range/Performance
P0787	Shift/Timing Solenoid Low
P0788	Shift/Timing Solenoid High
P0789	Shift/Timing Solenoid Intermittent
P0790	Normal/Performance Switch Circuit
P0791	Intermediate Shaft Speed Sensor "A" Circuit
P0792	Intermediate Shaft Speed Sensor "A" Circuit Range/Performance
P0793	Intermediate Shaft Speed Sensor "A" Circuit No Signal
P0794	Intermediate Shaft Speed Sensor "A" Circuit Intermittent
P0795	Pressure Control Solenoid "C"
P0796	Pressure Control Solenoid "C" Performance or Stuck off
P0797	Pressure Control Solenoid "C" Stuck On
P0798	Pressure Control Solenoid "C" Electrical
P0799	Pressure Control Solenoid "C" Intermittent
P0800	Transfer Case Control System (MIL Request)
P0801	Reverse Inhibit Control Circuit
P0802	Transmission Control System MIL Request Circuit/Open
P0803	1-4 Upshift (Skip Shift) Solenoid Control Circuit
P0804	1-4 Upshift (Skip Shift) Lamp Control Circuit
P0805	Clutch Position Sensor Circuit
P0806	Clutch Position Sensor Circuit Range/Performance
P0807	Clutch Position Sensor Circuit Low
P0808	Clutch Position Sensor Circuit High
P0809	Clutch Position Sensor Circuit Intermittent
P0810	Clutch Position Control Error
P0811	Excessive Clutch Slippage
P0812	Reverse Input Circuit
P0813	Reverse Output Circuit
P0814	Transmission Range Display Circuit
P0815	Upshift Switch Circuit
P0816	Downshift Switch Circuit

Vehicle CODE LOG

DATE	
Vehicle Info	
Vehicle CODE	P:
Current problems	
Solution	
NOTE	

P0817	Starter Disable Circuit
P0818	Driveline Disconnect Switch Input Circuit
P0819	Up and Down Shift Switch to Transmission Range Correlation
P0820	Gear Lever X-Y Position Sensor Circuit
P0821	Gear Lever X Position Circuit
P0822	Gear Lever Y Position Circuit
P0823	Gear Lever X Position Circuit Intermittent
P0824	Gear Lever Y Position Circuit Intermittent
P0825	Gear Lever Push-Pull Switch (Shift Anticipate)
P0826	Up and Down Shift Switch Circuit
P0827	Up and Down Shift Switch Circuit Low
P0828	Up and Down Shift Switch Circuit High
P0829	5-6 Shift
P0830	Clutch Pedal Switch "A" Circuit
P0831	Clutch Pedal Switch "A" Circuit Low
P0832	Clutch Pedal Switch "A" Circuit High
P0833	Clutch Pedal Switch "B" Circuit
P0834	Clutch Pedal Switch "B" Circuit Low
P0835	Clutch Pedal Switch "B" Circuit High
P0836	Four Wheel Drive (4WD) Switch Circuit
P0837	Four Wheel Drive (4WD) Switch Circuit Range/Performance
P0838	Four Wheel Drive (4WD) Switch Circuit Low
P0839	Four Wheel Drive (4WD) Switch Circuit High
P0840	Transmission Fluid Pressure Sensor/Switch "A" Circuit
P0841	Transmission Fluid Pressure Sensor/Switch "A" Circuit Range/Performance
P0842	Transmission Fluid Pressure Sensor/Switch "A" Circuit Low
P0843	Transmission Fluid Pressure Sensor/Switch "A" Circuit High
P0844	Transmission Fluid Pressure Sensor/Switch "A" Circuit Intermittent
P0845	Transmission Fluid Pressure Sensor/Switch "B" Circuit
P0846	Transmission Fluid Pressure Sensor/Switch "B" Circuit Range/Performance
P0847	Transmission Fluid Pressure Sensor/Switch "B" Circuit Low
P0848	Transmission Fluid Pressure Sensor/Switch "B" Circuit High
P0849	Transmission Fluid Pressure Sensor/Switch "B" Circuit Intermittent

Vehicle CODE LOG

DATE	
Vehicle Info	
Vehicle CODE	P:
Current problems	
Solution	
NOTE	

P0850	Park/Neutral Switch Input Circuit
P0851	Park/Neutral Switch Input Circuit Low
P0852	Park/Neutral Switch Input Circuit High
P0853	Drive Switch Input Circuit
P0854	Drive Switch Input Circuit Low
P0855	Drive Switch Input Circuit High
P0856	Traction Control Input Signal
P0857	Traction Control Input Signal Range/Performance
P0858	Traction Control Input Signal Low
P0859	Traction Control Input Signal High
P0860	Gear Shift Module Communication Circuit
P0861	Gear Shift Module Communication Circuit Low
P0862	Gear Shift Module Communication Circuit High
P0863	TCM Communication Circuit
P0864	TCM Communication Circuit Range/Performance
P0865	TCM Communication Circuit Low
P0866	TCM Communication Circuit High
P0867	Transmission Fluid Pressure
P0868	Transmission Fluid Pressure Low
P0869	Transmission Fluid Pressure High
P0870	Transmission Fluid Pressure Sensor/Switch "C" Circuit
P0871	Transmission Fluid Pressure Sensor/Switch "C" Circuit Range/Performance
P0872	Transmission Fluid Pressure Sensor/Switch "C" Circuit Low
P0873	Transmission Fluid Pressure Sensor/Switch "C" Circuit High
P0874	Transmission Fluid Pressure Sensor/Switch "C" Circuit Intermittent
P0875	Transmission Fluid Pressure Sensor/Switch "D" Circuit
P0876	Transmission Fluid Pressure Sensor/Switch "D" Circuit Range/Performance
P0877	Transmission Fluid Pressure Sensor/Switch "D" Circuit Low
P0878	Transmission Fluid Pressure Sensor/Switch "D" Circuit High
P0879	Transmission Fluid Pressure Sensor/Switch "D" Circuit Intermittent
P0880	TCM Power Input Signal
P0881	TCM Power Input Signal Range/Performance
P0882	TCM Power Input Signal Low

Vehicle CODE LOG

DATE	
Vehicle Info	
Vehicle CODE	P:
Current problems	
Solution	
NOTE	

P0883	TCM Power Input Signal High
P0884	TCM Power Input Signal Intermittent
P0885	TCM Power Relay Control Circuit/Open
P0886	TCM Power Relay Control Circuit Low
P0887	TCM Power Relay Control Circuit High
P0888	TCM Power Relay Sense Circuit
P0889	TCM Power Relay Sense Circuit Range/Performance
P0890	TCM Power Relay Sense Circuit Low
P0891	TCM Power Relay Sense Circuit High
P0892	TCM Power Relay Sense Circuit Intermittent
P0893	Multiple Gears Engaged
P0894	Transmission Component Slipping
P0895	Shift Time Too Short
P0896	Shift Time Too Long
P0897	Transmission Fluid Deteriorated
P0898	Transmission Control System MIL Request Circuit Low
P0899	Transmission Control System MIL Request Circuit High
P0900	Clutch Actuator Circuit/Open
P0901	Clutch Actuator Circuit Range/Performance
P0902	Clutch Actuator Circuit Low
P0903	Clutch Actuator Circuit High
P0904	Gate Select Position Circuit
P0905	Gate Select Position Circuit Range/Performance
P0906	Gate Select Position Circuit Low
P0907	Gate Select Position Circuit High
P0908	Gate Select Position Circuit Intermittent
P0909	Gate Select Control Error
P0910	Gate Select Actuator Circuit/Open
P0911	Gate Select Actuator Circuit Range/Performance
P0912	Gate Select Actuator Circuit Low
P0913	Gate Select Actuator Circuit High
P0914	Gear Shift Position Circuit
P0915	Gear Shift Position Circuit Range/Performance

Vehicle CODE LOG

DATE	
Vehicle Info	
Vehicle CODE	P:
Current problems	
Solution	
NOTE	

P0916	Gear Shift Position Circuit Low
P0917	Gear Shift Position Circuit High
P0918	Gear Shift Position Circuit Intermittent
P0919	Gear Shift Position Control Error
P0920	Gear Shift Forward Actuator Circuit/Open
P0921	Gear Shift Forward Actuator Circuit Range/Performance
P0922	Gear Shift Forward Actuator Circuit Low
P0923	Gear Shift Forward Actuator Circuit High
P0924	Gear Shift Reverse Actuator Circuit/Open
P0925	Gear Shift Reverse Actuator Circuit Range/Performance
P0926	Gear Shift Reverse Actuator Circuit Low
P0927	Gear Shift Reverse Actuator Circuit High
P0928	Gear Shift Lock Solenoid Control Circuit/Open
P0929	Gear Shift Lock Solenoid Control Circuit Range/Performance
P0930	Gear Shift Lock Solenoid Control Circuit Low
P0931	Gear Shift Lock Solenoid Control Circuit High
P0932	Hydraulic Pressure Sensor Circuit
P0933	Hydraulic Pressure Sensor Range/Performance
P0934	Hydraulic Pressure Sensor Circuit Low
P0935	Hydraulic Pressure Sensor Circuit High
P0936	Hydraulic Pressure Sensor Circuit Intermittent
P0937	Hydraulic Oil Temperature Sensor Circuit
P0938	Hydraulic Oil Temperature Sensor Range/Performance
P0939	Hydraulic Oil Temperature Sensor Circuit Low
P0940	Hydraulic Oil Temperature Sensor Circuit High
P0941	Hydraulic Oil Temperature Sensor Circuit Intermittent
P0942	Hydraulic Pressure Unit
P0943	Hydraulic Pressure Unit Cycling Period Too Short
P0944	Hydraulic Pressure Unit Loss of Pressure
P0945	Hydraulic Pump Relay Circuit/Open
P0946	Hydraulic Pump Relay Circuit Range/Performance
P0947	Hydraulic Pump Relay Circuit Low
P0948	Hydraulic Pump Relay Circuit High

Vehicle CODE LOG

DATE	
VehicleInfo	
VehicleCODE	P:
currentproblems	
Solution	
NOTE	

P0949	Auto Shift Manual Adaptive Learning Not Complete
P0950	Auto Shift Manual Control Circuit
P0951	Auto Shift Manual Control Circuit Range/Performance
P0952	Auto Shift Manual Control Circuit Low
P0953	Auto Shift Manual Control Circuit High
P0954	Auto Shift Manual Control Circuit Intermittent
P0955	Auto Shift Manual Mode Circuit
P0956	Auto Shift Manual Mode Circuit Range/Performance
P0957	Auto Shift Manual Mode Circuit Low
P0958	Auto Shift Manual Mode Circuit High
P0959	Auto Shift Manual Mode Circuit Intermittent
P0960	Pressure Control Solenoid "A" Control Circuit/Open
P0961	Pressure Control Solenoid "A" Control Circuit Range/Performance
P0962	Pressure Control Solenoid "A" Control Circuit Low
P0963	Pressure Control Solenoid "A" Control Circuit High
P0964	Pressure Control Solenoid "B" Control Circuit/Open
P0965	Pressure Control Solenoid "B" Control Circuit Range/Performance
P0966	Pressure Control Solenoid "B" Control Circuit Low
P0967	Pressure Control Solenoid "B" Control Circuit High
P0968	Pressure Control Solenoid "C" Control Circuit/Open
P0969	Pressure Control Solenoid "C" Control Circuit Range/Performance
P0970	Pressure Control Solenoid "C" Control Circuit Low
P0971	Pressure Control Solenoid "C" Control Circuit High
P0972	Shift Solenoid "A" Control Circuit Range/Performance
P0973	Shift Solenoid "A" Control Circuit Low
P0974	Shift Solenoid "A" Control Circuit High
P0975	Shift Solenoid "B" Control Circuit Range/Performance
P0976	Shift Solenoid "B" Control Circuit Low
P0977	Shift Solenoid "B" Control Circuit High
P0978	Shift Solenoid "C" Control Circuit Range/Performance
P0979	Shift Solenoid "C" Control Circuit Low
P0980	Shift Solenoid "C" Control Circuit High
P0981	Shift Solenoid "D" Control Circuit Range/Performance

Vehicle CODE LOG

DATE	
Vehicle Info	
Vehicle CODE	P:
Current problems	
Solution	
NOTE	

P0982	Shift Solenoid "D" Control Circuit Low
P0983	Shift Solenoid "D" Control Circuit High
P0984	Shift Solenoid "E" Control Circuit Range/Performance
P0985	Shift Solenoid "E" Control Circuit Low
P0986	Shift Solenoid "E" Control Circuit High
P0987	Transmission Fluid Pressure Sensor/Switch "E" Circuit
P0988	Transmission Fluid Pressure Sensor/Switch "E" Circuit Range/Performance
P0989	Transmission Fluid Pressure Sensor/Switch "E" Circuit Low
P0990	Transmission Fluid Pressure Sensor/Switch "E" Circuit High
P0991	Transmission Fluid Pressure Sensor/Switch "E" Circuit Intermittent
P0992	Transmission Fluid Pressure Sensor/Switch "F" Circuit
P0993	Transmission Fluid Pressure Sensor/Switch "F" Circuit Range/Performance
P0994	Transmission Fluid Pressure Sensor/Switch "F" Circuit Low
P0995	Transmission Fluid Pressure Sensor/Switch "F" Circuit High
P0996	Transmission Fluid Pressure Sensor/Switch "F" Circuit Intermittent
P0997	Shift Solenoid "F" Control Circuit Range/Performance
P0998	Shift Solenoid "F" Control Circuit Low
P0999	Shift Solenoid "F" Control Circuit High
P0A00	Motor Electronics Coolant Temperature Sensor Circuit
P0A01	Motor Electronics Coolant Temperature Sensor Circuit Range/Performance
P0A02	Motor Electronics Coolant Temperature Sensor Circuit Low
P0A03	Motor Electronics Coolant Temperature Sensor Circuit High
P0A04	Motor Electronics Coolant Temperature Sensor Circuit Intermittent
P0A05	Motor Electronics Coolant Pump Control Circuit/Open
P0A06	Motor Electronics Coolant Pump Control Circuit Low
P0A07	Motor Electronics Coolant Pump Control Circuit High
P0A08	DC/DC Converter Status Circuit
P0A09	DC/DC Converter Status Circuit Low Input
P0A10	DC/DC Converter Status Circuit High Input
P0A11	DC/DC Converter Enable Circuit/Open
P0A12	DC/DC Converter Enable Circuit Low
P0A13	DC/DC Converter Enable Circuit High
P0A14	Engine Mount Control Circuit/Open

Vehicle CODE LOG

DATE	
Vehicle Info	
Vehicle CODE	P:
Current problems	
Solution	
NOTE	

P0A15	Engine Mount Control Circuit Low
P0A16	Engine Mount Control Circuit High
P0A17	Motor Torque Sensor Circuit
P0A18	Motor Torque Sensor Circuit Range/Performance
P0A19	Motor Torque Sensor Circuit Low
P0A20	Motor Torque Sensor Circuit High
P0A21	Motor Torque Sensor Circuit Intermittent
P0A22	Generator Torque Sensor Circuit
P0A23	Generator Torque Sensor Circuit Range/Performance
P0A24	Generator Torque Sensor Circuit Low
P0A25	Generator Torque Sensor Circuit High
P0A26	Generator Torque Sensor Circuit Intermittent
P0A27	Battery Power Off Circuit
P0A28	Battery Power Off Circuit Low
P0A29	Battery Power Off Circuit High
P2000	NOx Trap Efficiency Below Threshold
P2001	NOx Trap Efficiency Below Threshold
P2002	Particulate Trap Efficiency Below Threshold
P2003	Particulate Trap Efficiency Below Threshold
P2004	Intake Manifold Runner Control Stuck Open
P2005	Intake Manifold Runner Control Stuck Open
P2006	Intake Manifold Runner Control Stuck Closed
P2007	Intake Manifold Runner Control Stuck Closed
P2008	Intake Manifold Runner Control Circuit/Open
P2009	Intake Manifold Runner Control Circuit Low
P2010	Intake Manifold Runner Control Circuit High
P2011	Intake Manifold Runner Control Circuit/Open
P2012	Intake Manifold Runner Control Circuit Low
P2013	Intake Manifold Runner Control Circuit High
P2014	Intake Manifold Runner Position Sensor/Switch Circuit
P2015	Intake Manifold Runner Position Sensor/Switch Circuit Range/Performance
P2016	Intake Manifold Runner Position Sensor/Switch Circuit Low
P2017	Intake Manifold Runner Position Sensor/Switch Circuit High

Vehicle CODE LOG

DATE	
Vehicle Info	
Vehicle CODE	P:
Current problems	
Solution	
NOTE	

P2018	Intake Manifold Runner Position Sensor/Switch Circuit Intermittent
P2019	Intake Manifold Runner Position Sensor/Switch Circuit
P2020	Intake Manifold Runner Position Sensor/Switch Circuit Range/Performance
P2021	Intake Manifold Runner Position Sensor/Switch Circuit Low
P2022	Intake Manifold Runner Position Sensor/Switch Circuit High
P2023	Intake Manifold Runner Position Sensor/Switch Circuit Intermittent
P2024	Evaporative Emissions (EVAP) Fuel Vapor Temperature Sensor Circuit
P2025	Evaporative Emissions (EVAP) Fuel Vapor Temperature Sensor Performance
P2026	Evaporative Emissions (EVAP) Fuel Vapor Temperature Sensor Circuit Low Voltage
P2027	Evaporative Emissions (EVAP) Fuel Vapor Temperature Sensor Circuit High Voltage
P2028	Evaporative Emissions (EVAP) Fuel Vapor Temperature Sensor Circuit Intermittent
P2029	Fuel Fired Heater Disabled
P2030	Fuel Fired Heater Performance
P2031	Exhaust Gas Temperature Sensor Circuit
P2032	Exhaust Gas Temperature Sensor Circuit Low
P2033	Exhaust Gas Temperature Sensor Circuit High
P2034	Exhaust Gas Temperature Sensor Circuit
P2035	Exhaust Gas Temperature Sensor Circuit Low
P2036	Exhaust Gas Temperature Sensor Circuit High
P2037	Reductant Injection Air Pressure Sensor Circuit
P2038	Reductant Injection Air Pressure Sensor Circuit Range/Performance
P2039	Reductant Injection Air Pressure Sensor Circuit Low Input
P2040	Reductant Injection Air Pressure Sensor Circuit High Input
P2041	Reductant Injection Air Pressure Sensor Circuit Intermittent
P2042	Reductant Temperature Sensor Circuit
P2043	Reductant Temperature Sensor Circuit Range/Performance
P2044	Reductant Temperature Sensor Circuit Low Input
P2045	Reductant Temperature Sensor Circuit High Input
P2046	Reductant Temperature Sensor Circuit Intermittent
P2047	Reductant Injector Circuit/Open
P2048	Reductant Injector Circuit Low
P2049	Reductant Injector Circuit High
P2050	Reductant Injector Circuit/Open

Vehicle CODE LOG

DATE	
Vehicle Info	
Vehicle CODE	P:
Current problems	
Solution	
NOTE	

P2051	Reductant Injector Circuit Low
P2052	Reductant Injector Circuit High
P2053	Reductant Injector Circuit/Open
P2054	Reductant Injector Circuit Low
P2055	Reductant Injector Circuit High
P2056	Reductant Injector Circuit/Open
P2057	Reductant Injector Circuit Low
P2058	Reductant Injector Circuit High
P2059	Reductant Injection Air Pump Control Circuit/Open
P2060	Reductant Injection Air Pump Control Circuit Low
P2061	Reductant Injection Air Pump Control Circuit High
P2062	Reductant Supply Control Circuit/Open
P2063	Reductant Supply Control Circuit Low
P2064	Reductant Supply Control Circuit High
P2065	Fuel Level Sensor "B" Circuit
P2066	Fuel Level Sensor "B" Performance
P2067	Fuel Level Sensor "B" Circuit Low
P2068	Fuel Level Sensor "B" Circuit High
P2069	Fuel Level Sensor "B" Circuit Intermittent
P2070	Intake Manifold Tuning (IMT) Valve Stuck Open
P2071	Intake Manifold Tuning (IMT) Valve Stuck Closed
P2075	Intake Manifold Tuning (IMT) Valve Position Sensor/Switch Circuit
P2076	Intake Manifold Tuning (IMT) Valve Position Sensor/Switch Circuit Range/Performance
P2077	Intake Manifold Tuning (IMT) Valve Position Sensor/Switch Circuit Low
P2078	Intake Manifold Tuning (IMT) Valve Position Sensor/Switch Circuit High
P2079	Intake Manifold Tuning (IMT) Valve Position Sensor/Switch Circuit Intermittent
P2080	Exhaust Gas Temperature Sensor Circuit Range/Performance
P2081	Exhaust Gas Temperature Sensor Circuit Intermittent
P2082	Exhaust Gas Temperature Sensor Circuit Range/Performance
P2083	Exhaust Gas Temperature Sensor Circuit Intermittent
P2084	Exhaust Gas Temperature Sensor Circuit Range/Performance
P2085	Exhaust Gas Temperature Sensor Circuit Intermittent
P2086	Exhaust Gas Temperature Sensor Circuit Range/Performance

Vehicle CODE LOG

DATE	
Vehicle Info	
Vehicle CODE	P:
Current problems	
Solution	
NOTE	

P2087	Exhaust Gas Temperature Sensor Circuit Intermittent
P2088	"A" Camshaft Position Actuator Control Circuit Low
P2089	"A" Camshaft Position Actuator Control Circuit High
P2090	"B" Camshaft Position Actuator Control Circuit Low
P2091	"B" Camshaft Position Actuator Control Circuit High
P2092	"A" Camshaft Position Actuator Control Circuit Low
P2093	"A" Camshaft Position Actuator Control Circuit High
P2094	"B" Camshaft Position Actuator Control Circuit Low
P2095	"B" Camshaft Position Actuator Control Circuit High
P2096	Post Catalyst Fuel Trim System Too Lean
P2097	Post Catalyst Fuel Trim System Too Rich
P2098	Post Catalyst Fuel Trim System Too Lean
P2099	Post Catalyst Fuel Trim System Too Rich
P2100	Throttle Actuator Control Motor Circuit/Open
P2101	Throttle Actuator Control Motor Circuit Range/Performance
P2102	Throttle Actuator Control Motor Circuit Low
P2103	Throttle Actuator Control Motor Circuit High
P2104	Throttle Actuator Control System - Forced Idle
P2105	Throttle Actuator Control System - Forced Engine Shutdown
P2106	Throttle Actuator Control System - Forced Limited Power
P2107	Throttle Actuator Control Module Processor
P2108	Throttle Actuator Control Module Performance
P2109	Throttle/Pedal Position Sensor "A" Minimum Stop Performance
P2110	Throttle Actuator Control System - Forced Limited RPM
P2111	Throttle Actuator Control System - Stuck Open
P2112	Throttle Actuator Control System - Stuck Closed
P2113	Throttle/Pedal Position Sensor "B" Minimum Stop Performance
P2114	Throttle/Pedal Position Sensor "C" Minimum Stop Performance
P2115	Throttle/Pedal Position Sensor "D" Minimum Stop Performance
P2116	Throttle/Pedal Position Sensor "E" Minimum Stop Performance
P2117	Throttle/Pedal Position Sensor "F" Minimum Stop Performance
P2118	Throttle Actuator Control Motor Current Range/Performance
P2119	Throttle Actuator Control Throttle Body Range/Performance

Vehicle CODE LOG

DATE	
Vehicle Info	
Vehicle CODE	P:
Current problems	
Solution	
NOTE	

P2120	Throttle/Pedal Position Sensor/Switch "D" Circuit
P2121	Throttle/Pedal Position Sensor/Switch "D" Circuit Range/Performance
P2122	Throttle/Pedal Position Sensor/Switch "D" Circuit Low Input
P2123	Throttle/Pedal Position Sensor/Switch "D" Circuit High Input
P2124	Throttle/Pedal Position Sensor/Switch "D" Circuit Intermittent
P2125	Throttle/Pedal Position Sensor/Switch "E" Circuit
P2126	Throttle/Pedal Position Sensor/Switch "E" Circuit Range/Performance
P2127	Throttle/Pedal Position Sensor/Switch "E" Circuit Low Input
P2128	Throttle/Pedal Position Sensor/Switch "E" Circuit High Input
P2129	Throttle/Pedal Position Sensor/Switch "E" Circuit Intermittent
P2130	Throttle/Pedal Position Sensor/Switch "F" Circuit
P2131	Throttle/Pedal Position Sensor/Switch "F" Circuit Range Performance
P2132	Throttle/Pedal Position Sensor/Switch "F" Circuit Low Input
P2133	Throttle/Pedal Position Sensor/Switch "F" Circuit High Input
P2134	Throttle/Pedal Position Sensor/Switch "F" Circuit Intermittent
P2135	Throttle/Pedal Position Sensor/Switch "A" / "B" Voltage Correlation
P2136	Throttle/Pedal Position Sensor/Switch "A" / "C" Voltage Correlation
P2137	Throttle/Pedal Position Sensor/Switch "B" / "C" Voltage Correlation
P2138	Throttle/Pedal Position Sensor/Switch "D" / "E" Voltage Correlation
P2139	Throttle/Pedal Position Sensor/Switch "D" / "F" Voltage Correlation
P2140	Throttle/Pedal Position Sensor/Switch "E" / "F" Voltage Correlation
P2141	Exhaust Gas Recirculation Throttle Control Circuit Low
P2142	Exhaust Gas Recirculation Throttle Control Circuit High
P2143	Exhaust Gas Recirculation Vent Control Circuit/Open
P2144	Exhaust Gas Recirculation Vent Control Circuit Low
P2145	Exhaust Gas Recirculation Vent Control Circuit High
P2146	Fuel Injector Group "A" Supply Voltage Circuit/Open
P2147	Fuel Injector Group "A" Supply Voltage Circuit Low
P2148	Fuel Injector Group "A" Supply Voltage Circuit High
P2149	Fuel Injector Group "B" Supply Voltage Circuit/Open
P2150	Fuel Injector Group "B" Supply Voltage Circuit Low
P2151	Fuel Injector Group "B" Supply Voltage Circuit High
P2152	Fuel Injector Group "C" Supply Voltage Circuit/Open

Vehicle CODE LOG

DATE	
Vehicle Info	
Vehicle CODE	P:
Current problems	
Solution	
NOTE	

P2153	Fuel Injector Group "C" Supply Voltage Circuit Low
P2154	Fuel Injector Group "C" Supply Voltage Circuit High
P2155	Fuel Injector Group "D" Supply Voltage Circuit/Open
P2156	Fuel Injector Group "D" Supply Voltage Circuit Low
P2157	Fuel Injector Group "D" Supply Voltage Circuit High
P2158	Vehicle Speed Sensor "B"
P2159	Vehicle Speed Sensor "B" Range/Performance
P2160	Vehicle Speed Sensor "B" Circuit Low
P2161	Vehicle Speed Sensor "B" Intermittent/Erratic
P2162	Vehicle Speed Sensor "A" / "B" Correlation
P2163	Throttle/Pedal Position Sensor "A" Maximum Stop Performance
P2164	Throttle/Pedal Position Sensor "B" Maximum Stop Performance
P2165	Throttle/Pedal Position Sensor "C" Maximum Stop Performance
P2166	Throttle/Pedal Position Sensor "D" Maximum Stop Performance
P2167	Throttle/Pedal Position Sensor "E" Maximum Stop Performance
P2168	Throttle/Pedal Position Sensor "F" Maximum Stop Performance
P2169	Exhaust Pressure Regulator Vent Solenoid Control Circuit/Open
P2170	Exhaust Pressure Regulator Vent Solenoid Control Circuit Low
P2171	Exhaust Pressure Regulator Vent Solenoid Control Circuit High
P2172	Throttle Actuator Control System - Sudden High Airflow Detected
P2173	Throttle Actuator Control System - High Airflow Detected
P2174	Throttle Actuator Control System - Sudden Low Airflow Detected
P2175	Throttle Actuator Control System - Low Airflow Detected
P2176	Throttle Actuator Control System - Idle Position Not Learned
P2177	System Too Lean Off Idle
P2178	System Too Rich Off Idle
P2179	System Too Lean Off Idle
P2180	System Too Rich Off Idle
P2181	Cooling System Performance
P2182	Engine Coolant Temperature Sensor 2 Circuit
P2183	Engine Coolant Temperature Sensor 2 Circuit Range/Performance
P2184	Engine Coolant Temperature Sensor 2 Circuit Low
P2185	Engine Coolant Temperature Sensor 2 Circuit High

Vehicle CODE LOG

DATE	
Vehicle Info	
Vehicle CODE	P:
Current problems	
Solution	
NOTE	

P2186	Engine Coolant Temperature Sensor 2 Circuit Intermittent/Erratic
P2187	System Too Lean at Idle
P2188	System Too Rich at Idle
P2189	System Too Lean at Idle
P2190	System Too Rich at Idle
P2191	System Too Lean at Higher Load
P2192	System Too Rich at Higher Load
P2193	System Too Lean at Higher Load
P2194	System Too Rich at Higher Load
P2195	O2 Sensor Signal Stuck Lean
P2196	O2 Sensor Signal Stuck Rich
P2197	O2 Sensor Signal Stuck Lean
P2198	O2 Sensor Signal Stuck Rich
P2199	Intake Air Temperature Sensor 1 / 2 Correlation
P2200	NOx Sensor Circuit
P2201	NOx Sensor Circuit Range/Performance
P2202	NOx Sensor Circuit Low Input
P2203	NOx Sensor Circuit High Input
P2204	NOx Sensor Circuit Intermittent Input
P2205	NOx Sensor Heater Control Circuit/Open
P2206	NOx Sensor Heater Control Circuit Low
P2207	NOx Sensor Heater Control Circuit High
P2208	NOx Sensor Heater Sense Circuit
P2209	NOx Sensor Heater Sense Circuit Range/Performance
P2210	NOx Sensor Heater Sense Circuit Low Input
P2211	NOx Sensor Heater Sense Circuit High Input
P2212	NOx Sensor Heater Sense Circuit Intermittent
P2213	NOx Sensor Circuit
P2214	NOx Sensor Circuit Range/Performance
P2215	NOx Sensor Circuit Low Input
P2216	NOx Sensor Circuit High Input
P2217	NOx Sensor Circuit Intermittent Input
P2218	NOx Sensor Heater Control Circuit/Open

Vehicle CODE LOG

DATE	
Vehicle Info	
Vehicle CODE	P:
Current problems	
Solution	
NOTE	

P2219	NOx Sensor Heater Control Circuit Low
P2220	NOx Sensor Heater Control Circuit High
P2221	NOx Sensor Heater Sense Circuit
P2222	NOx Sensor Heater Sense Circuit Range/Performance
P2223	NOx Sensor Heater Sense Circuit Low
P2224	NOx Sensor Heater Sense Circuit High
P2225	NOx Sensor Heater Sense Circuit Intermittent
P2226	Barometric Pressure Circuit
P2227	Barometric Pressure Circuit Range/Performance
P2228	Barometric Pressure Circuit Low
P2229	Barometric Pressure Circuit High
P2230	Barometric Pressure Circuit Intermittent
P2231	O2 Sensor Signal Circuit Shorted to Heater Circuit
P2232	O2 Sensor Signal Circuit Shorted to Heater Circuit
P2233	O2 Sensor Signal Circuit Shorted to Heater Circuit
P2234	O2 Sensor Signal Circuit Shorted to Heater Circuit
P2235	O2 Sensor Signal Circuit Shorted to Heater Circuit
P2236	O2 Sensor Signal Circuit Shorted to Heater Circuit
P2237	O2 Sensor Positive Current Control Circuit/Open
P2238	O2 Sensor Positive Current Control Circuit Low
P2239	O2 Sensor Positive Current Control Circuit High
P2240	O2 Sensor Positive Current Control Circuit/Open
P2241	O2 Sensor Positive Current Control Circuit Low
P2242	O2 Sensor Positive Current Control Circuit High
P2243	O2 Sensor Reference Voltage Circuit/Open
P2244	O2 Sensor Reference Voltage Performance
P2245	O2 Sensor Reference Voltage Circuit Low
P2246	O2 Sensor Reference Voltage Circuit High
P2247	O2 Sensor Reference Voltage Circuit/Open
P2248	O2 Sensor Reference Voltage Performance
P2249	O2 Sensor Reference Voltage Circuit Low
P2250	O2 Sensor Reference Voltage Circuit High
P2251	O2 Sensor Negative Current Control Circuit/Open

Vehicle CODE LOG

DATE	
Vehicle Info	
Vehicle CODE	P:
Current problems	
Solution	
NOTE	

P2252	O2 Sensor Negative Current Control Circuit Low
P2253	O2 Sensor Negative Current Control Circuit High
P2254	O2 Sensor Negative Current Control Circuit/Open
P2255	O2 Sensor Negative Current Control Circuit Low
P2256	O2 Sensor Negative Current Control Circuit High
P2257	Secondary Air Injection System Control "A" Circuit Low
P2258	Secondary Air Injection System Control "A" Circuit High
P2259	Secondary Air Injection System Control "B" Circuit Low
P2260	Secondary Air Injection System Control "B" Circuit High
P2261	Turbo/Super Charger Bypass Valve - Mechanical
P2262	Turbo Boost Pressure Not Detected - Mechanical
P2263	Turbo/Super Charger Boost System Performance
P2264	Water in Fuel Sensor Circuit
P2265	Water in Fuel Sensor Circuit Range/Performance
P2266	Water in Fuel Sensor Circuit Low
P2267	Water in Fuel Sensor Circuit High
P2268	Water in Fuel Sensor Circuit Intermittent
P2269	Water in Fuel Condition
P2270	O2 Sensor Signal Stuck Lean
P2271	O2 Sensor Signal Stuck Rich
P2272	O2 Sensor Signal Stuck Lean
P2273	O2 Sensor Signal Stuck Rich
P2274	O2 Sensor Signal Stuck Lean
P2275	O2 Sensor Signal Stuck Rich
P2276	O2 Sensor Signal Stuck Lean
P2277	O2 Sensor Signal Stuck Rich
P2278	O2 Sensor Signals Swapped Bank 1 Sensor 3 / Bank 2 Sensor 3
P2279	Intake Air System Leak
P2280	Air Flow Restriction / Air Leak Between Air Filter and MAF
P2281	Air Leak Between MAF and Throttle Body
P2282	Air Leak Between Throttle Body and Intake Valves
P2283	Injector Control Pressure Sensor Circuit
P2284	Injector Control Pressure Sensor Circuit Range/Performance

VehicleCODELOG

DATE	
Vehicle Info	
Vehicle CODE	P:
Current problems	
Solution	
NOTE	

P2285	Injector Control Pressure Sensor Circuit Low
P2286	Injector Control Pressure Sensor Circuit High
P2287	Injector Control Pressure Sensor Circuit Intermittent
P2288	Injector Control Pressure Too High
P2289	Injector Control Pressure Too High - Engine Off
P2290	Injector Control Pressure Too Low
P2291	Injector Control Pressure Too Low - Engine Cranking
P2292	Injector Control Pressure Erratic
P2293	Fuel Pressure Regulator 2 Performance
P2294	Fuel Pressure Regulator 2 Control Circuit
P2295	Fuel Pressure Regulator 2 Control Circuit Low
P2296	Fuel Pressure Regulator 2 Control Circuit High
P2297	O2 Sensor Out of Range During Deceleration
P2298	O2 Sensor Out of Range During Deceleration
P2299	Brake Pedal Position / Accelerator Pedal Position Incompatible
P2300	Ignition Coil "A" Primary Control Circuit Low
P2301	Ignition Coil "A" Primary Control Circuit High
P2302	Ignition Coil "A" Secondary Circuit
P2303	Ignition Coil "B" Primary Control Circuit Low
P2304	Ignition Coil "B" Primary Control Circuit High
P2305	Ignition Coil "B" Secondary Circuit
P2306	Ignition Coil "C" Primary Control Circuit Low
P2307	Ignition Coil "C" Primary Control Circuit High
P2308	Ignition Coil "C" Secondary Circuit
P2309	Ignition Coil "D" Primary Control Circuit Low
P2310	Ignition Coil "D" Primary Control Circuit High
P2311	Ignition Coil "D" Secondary Circuit
P2312	Ignition Coil "E" Primary Control Circuit Low
P2313	Ignition Coil "E" Primary Control Circuit High
P2314	Ignition Coil "E" Secondary Circuit
P2315	Ignition Coil "F" Primary Control Circuit Low
P2316	Ignition Coil "F" Primary Control Circuit High
P2317	Ignition Coil "F" Secondary Circuit

Vehicle CODE LOG

DATE	
Vehicle Info	
Vehicle CODE	P:
Current problems	
Solution	
NOTE	

P2318	Ignition Coil "G" Primary Control Circuit Low
P2319	Ignition Coil "G" Primary Control Circuit High
P2320	Ignition Coil "G" Secondary Circuit
P2321	Ignition Coil "H" Primary Control Circuit Low
P2322	Ignition Coil "H" Primary Control Circuit High
P2323	Ignition Coil "H" Secondary Circuit
P2324	Ignition Coil "I" Primary Control Circuit Low
P2325	Ignition Coil "I" Primary Control Circuit High
P2326	Ignition Coil "I" Secondary Circuit
P2327	Ignition Coil "J" Primary Control Circuit Low
P2328	Ignition Coil "J" Primary Control Circuit High
P2329	Ignition Coil "J" Secondary Circuit
P2330	Ignition Coil "K" Primary Control Circuit Low
P2331	Ignition Coil "K" Primary Control Circuit High
P2332	Ignition Coil "K" Secondary Circuit
P2333	Ignition Coil "L" Primary Control Circuit Low
P2334	Ignition Coil "L" Primary Control Circuit High
P2335	Ignition Coil "L" Secondary Circuit
P2336	Cylinder #1 Above Knock Threshold
P2337	Cylinder #2 Above Knock Threshold
P2338	Cylinder #3 Above Knock Threshold
P2339	Cylinder #4 Above Knock Threshold
P2340	Cylinder #5 Above Knock Threshold
P2341	Cylinder #6 Above Knock Threshold
P2342	Cylinder #7 Above Knock Threshold
P2343	Cylinder #8 Above Knock Threshold
P2344	Cylinder #9 Above Knock Threshold
P2345	Cylinder #10 Above Knock Threshold
P2346	Cylinder #11 Above Knock Threshold
P2347	Cylinder #12 Above Knock Threshold
P2400	Evaporative Emission System Leak Detection Pump Control Circuit/Open
P2401	Evaporative Emission System Leak Detection Pump Control Circuit Low
P2402	Evaporative Emission System Leak Detection Pump Control Circuit High

Vehicle CODE LOG

DATE	
Vehicle Info	
Vehicle CODE	P:
Current problems	
Solution	
NOTE	

P2403	Evaporative Emission System Leak Detection Pump Sense Circuit/Open
P2404	Evaporative Emission System Leak Detection Pump Sense Circuit Range/Performance
P2405	Evaporative Emission System Leak Detection Pump Sense Circuit Low
P2406	Evaporative Emission System Leak Detection Pump Sense Circuit High
P2407	Evaporative Emission System Leak Detection Pump Sense Circuit Intermittent/Erratic
P2408	Fuel Cap Sensor/Switch Circuit
P2409	Fuel Cap Sensor/Switch Circuit Range/Performance
P2410	Fuel Cap Sensor/Switch Circuit Low
P2411	Fuel Cap Sensor/Switch Circuit High
P2412	Fuel Cap Sensor/Switch Circuit Intermittent/Erratic
P2413	Exhaust Gas Recirculation System Performance
P2414	O2 Sensor Exhaust Sample Error
P2415	O2 Sensor Exhaust Sample Error
P2416	O2 Sensor Signals Swapped Bank 1 Sensor 2 / Bank 1 Sensor 3
P2417	O2 Sensor Signals Swapped Bank 2 Sensor 2 / Bank 2 Sensor 3
P2418	Evaporative Emission System Switching Valve Control Circuit / Open
P2419	Evaporative Emission System Switching Valve Control Circuit Low
P2420	Evaporative Emission System Switching Valve Control Circuit High
P2421	Evaporative Emission System Vent Valve Stuck Open
P2422	Evaporative Emission System Vent Valve Stuck Closed
P2423	HC Adsorption Catalyst Efficiency Below Threshold
P2424	HC Adsorption Catalyst Efficiency Below Threshold
P2425	Exhaust Gas Recirculation Cooling Valve Control Circuit/Open
P2426	Exhaust Gas Recirculation Cooling Valve Control Circuit Low
P2427	Exhaust Gas Recirculation Cooling Valve Control Circuit High
P2428	Exhaust Gas Temperature Too High
P2429	Exhaust Gas Temperature Too High
P2430	Secondary Air Injection System Air Flow/Pressure Sensor Circuit
P2431	Secondary Air Injection System Air Flow/Pressure Sensor Circuit Range/Performance
P2432	Secondary Air Injection System Air Flow/Pressure Sensor Circuit Low
P2433	Secondary Air Injection System Air Flow/Pressure Sensor Circuit High
P2434	Secondary Air Injection System Air Flow/Pressure Sensor Circuit Intermittent/Erratic
P2435	Secondary Air Injection System Air Flow/Pressure Sensor Circuit

Vehicle CODE LOG

DATE	
Vehicle Info	
Vehicle CODE	P:
Current problems	
Solution	
NOTE	

P2436	Secondary Air Injection System Air Flow/Pressure Sensor Circuit Range/Performance
P2437	Secondary Air Injection System Air Flow/Pressure Sensor Circuit Low
P2438	Secondary Air Injection System Air Flow/Pressure Sensor Circuit High
P2439	Secondary Air Injection System Air Flow/Pressure Sensor Circuit Intermittent/Erratic
P2440	Secondary Air Injection System Switching Valve Stuck Open
P2441	Secondary Air Injection System Switching Valve Stuck Closed
P2442	Secondary Air Injection System Switching Valve Stuck Open
P2443	Secondary Air Injection System Switching Valve Stuck Closed
P2444	Secondary Air Injection System Pump Stuck On
P2445	Secondary Air Injection System Pump Stuck Off
P2446	Secondary Air Injection System Pump Stuck On
P2447	Secondary Air Injection System Pump Stuck Off
P2500	Generator Lamp/L-Terminal Circuit Low
P2501	Generator Lamp/L-Terminal Circuit High
P2502	Charging System Voltage
P2503	Charging System Voltage Low
P2504	Charging System Voltage High
P2505	ECM/PCM Power Input Signal
P2506	ECM/PCM Power Input Signal Range/Performance
P2507	ECM/PCM Power Input Signal Low
P2508	ECM/PCM Power Input Signal High
P2509	ECM/PCM Power Input Signal Intermittent
P2510	ECM/PCM Power Relay Sense Circuit Range/Performance
P2511	ECM/PCM Power Relay Sense Circuit Intermittent
P2512	Event Data Recorder Request Circuit/ Open
P2513	Event Data Recorder Request Circuit Low
P2514	Event Data Recorder Request Circuit High
P2515	A/C Refrigerant Pressure Sensor "B" Circuit
P2516	A/C Refrigerant Pressure Sensor "B" Circuit Range/Performance
P2517	A/C Refrigerant Pressure Sensor "B" Circuit Low
P2518	A/C Refrigerant Pressure Sensor "B" Circuit High
P2519	A/C Request "A" Circuit
P2520	A/C Request "A" Circuit Low

Vehicle CODE LOG

DATE	
Vehicle Info	
Vehicle CODE	P:
Current problems	
Solution	
NOTE	

P2521	A/C Request "A" Circuit High
P2522	A/C Request "B" Circuit
P2523	A/C Request "B" Circuit Low
P2524	A/C Request "B" Circuit High
P2525	Vacuum Reservoir Pressure Sensor Circuit
P2526	Vacuum Reservoir Pressure Sensor Circuit Range/Performance
P2527	Vacuum Reservoir Pressure Sensor Circuit Low
P2528	Vacuum Reservoir Pressure Sensor Circuit High
P2529	Vacuum Reservoir Pressure Sensor Circuit Intermittent
P2530	Ignition Switch Run Position Circuit
P2531	Ignition Switch Run Position Circuit Low
P2532	Ignition Switch Run Position Circuit High
P2533	Ignition Switch Run/Start Position Circuit
P2534	Ignition Switch Run/Start Position Circuit Low
P2535	Ignition Switch Run/Start Position Circuit High
P2536	Ignition Switch Accessory Position Circuit
P2537	Ignition Switch Accessory Position Circuit Low
P2538	Ignition Switch Accessory Position Circuit High
P2539	Low Pressure Fuel System Sensor Circuit
P2540	Low Pressure Fuel System Sensor Circuit Range/Performance
P2541	Low Pressure Fuel System Sensor Circuit Low
P2542	Low Pressure Fuel System Sensor Circuit High
P2543	Low Pressure Fuel System Sensor Circuit Intermittent
P2544	Torque Management Request Input Signal "A"
P2545	Torque Management Request Input Signal "A" Range/Performance
P2546	Torque Management Request Input Signal "A" Low
P2547	Torque Management Request Input Signal "A" High
P2548	Torque Management Request Input Signal "B"
P2549	Torque Management Request Input Signal "B" Range/Performance
P2550	Torque Management Request Input Signal "B" Low
P2551	Torque Management Request Input Signal "B" High
P2552	Throttle/Fuel Inhibit Circuit
P2553	Throttle/Fuel Inhibit Circuit Range/Performance

Vehicle CODE LOG

DATE	
Vehicle Info	
Vehicle CODE	P:
Current problems	
Solution	
NOTE	

P2554	Throttle/Fuel Inhibit Circuit Low
P2555	Throttle/Fuel Inhibit Circuit High
P2556	Engine Coolant Level Sensor/Switch Circuit
P2557	Engine Coolant Level Sensor/Switch Circuit Range/Performance
P2558	Engine Coolant Level Sensor/Switch Circuit Low
P2559	Engine Coolant Level Sensor/Switch Circuit High
P2560	Engine Coolant Level Low
P2561	A/C Control Module Requested MIL Illumination
P2562	Turbocharger Boost Control Position Sensor Circuit
P2563	Turbocharger Boost Control Position Sensor Circuit Range/Performance
P2564	Turbocharger Boost Control Position Sensor Circuit Low
P2565	Turbocharger Boost Control Position Sensor Circuit High
P2566	Turbocharger Boost Control Position Sensor Circuit Intermittent
P2567	Direct Ozone Reduction Catalyst Temperature Sensor Circuit
P2568	Direct Ozone Reduction Catalyst Temperature Sensor Circuit Range/Performance
P2569	Direct Ozone Reduction Catalyst Temperature Sensor Circuit Low
P2570	Direct Ozone Reduction Catalyst Temperature Sensor Circuit High
P2571	Direct Ozone Reduction Catalyst Temperature Sensor Circuit Intermittent/Erratic
P2572	Direct Ozone Reduction Catalyst Deterioration Sensor Circuit
P2573	Direct Ozone Reduction Catalyst Deterioration Sensor Circuit Range/Performance
P2574	Direct Ozone Reduction Catalyst Deterioration Sensor Circuit Low
P2575	Direct Ozone Reduction Catalyst Deterioration Sensor Circuit High
P2576	Direct Ozone Reduction Catalyst Deterioration Sensor Circuit Intermittent/Erratic
P2577	Direct Ozone Reduction Catalyst Efficiency Below Threshold
P2600	Coolant Pump Control Circuit/Open
P2601	Coolant Pump Control Circuit Range/Performance
P2602	Coolant Pump Control Circuit Low
P2603	Coolant Pump Control Circuit High
P2604	Intake Air Heater "A" Circuit Range/Performance
P2605	Intake Air Heater "A" Circuit/Open
P2606	Intake Air Heater "B" Circuit Range/Performance
P2607	Intake Air Heater "B" Circuit Low
P2608	Intake Air Heater "B" Circuit High

Vehicle CODE LOG

DATE	
Vehicle Info	
Vehicle CODE	P:
Current problems	
Solution	
NOTE	

P2609	Intake Air Heater System Performance
P2610	ECM/PCM Internal Engine Off Timer Performance
P2611	A/C Refrigerant Distribution Valve Control Circuit/Open
P2612	A/C Refrigerant Distribution Valve Control Circuit Low
P2613	A/C Refrigerant Distribution Valve Control Circuit High
P2614	Camshaft Position Signal Output Circuit/Open
P2615	Camshaft Position Signal Output Circuit Low
P2616	Camshaft Position Signal Output Circuit High
P2617	Crankshaft Position Signal Output Circuit/Open
P2618	Crankshaft Position Signal Output Circuit Low
P2619	Crankshaft Position Signal Output Circuit High
P2620	Throttle Position Output Circuit/Open
P2621	Throttle Position Output Circuit Low
P2622	Throttle Position Output Circuit High
P2623	Injector Control Pressure Regulator Circuit/Open
P2624	Injector Control Pressure Regulator Circuit Low
P2625	Injector Control Pressure Regulator Circuit High
P2626	O2 Sensor Pumping Current Trim Circuit/Open
P2627	O2 Sensor Pumping Current Trim Circuit Low
P2628	O2 Sensor Pumping Current Trim Circuit High
P2629	O2 Sensor Pumping Current Trim Circuit/Open
P2630	O2 Sensor Pumping Current Trim Circuit Low
P2631	O2 Sensor Pumping Current Trim Circuit High
P2632	Fuel Pump "B" Control Circuit /Open
P2633	Fuel Pump "B" Control Circuit Low
P2634	Fuel Pump "B" Control Circuit High
P2635	Fuel Pump "A" Low Flow / Performance
P2636	Fuel Pump "B" Low Flow / Performance
P2637	Torque Management Feedback Signal "A"
P2638	Torque Management Feedback Signal "A" Range/Performance
P2639	Torque Management Feedback Signal "A" Low
P2640	Torque Management Feedback Signal "A" High
P2641	Torque Management Feedback Signal "B"

Vehicle CODE LOG

DATE	
Vehicle Info	
Vehicle CODE	P:
Current problems	
Solution	
NOTE	

P2642	Torque Management Feedback Signal "B" Range/Performance
P2643	Torque Management Feedback Signal "B" Low
P2644	Torque Management Feedback Signal "B" High
P2645	"A" Rocker Arm Actuator Control Circuit/Open
P2646	"A" Rocker Arm Actuator System Performance or Stuck Off
P2647	"A" Rocker Arm Actuator System Stuck On
P2648	"A" Rocker Arm Actuator Control Circuit Low
P2649	"A" Rocker Arm Actuator Control Circuit High
P2650	"B" Rocker Arm Actuator Control Circuit/Open
P2651	"B" Rocker Arm Actuator System Performance or Stuck Off
P2652	"B" Rocker Arm Actuator System Stuck On
P2653	"B" Rocker Arm Actuator Control Circuit Low
P2654	"B" Rocker Arm Actuator Control Circuit High
P2655	"A" Rocker Arm Actuator Control Circuit/Open
P2656	"A" Rocker Arm Actuator System Performance or Stuck Off
P2657	"A" Rocker Arm Actuator System Stuck On
P2658	"A" Rocker Arm Actuator Control Circuit Low
P2659	"A" Rocker Arm Actuator Control Circuit High
P2660	"B" Rocker Arm Actuator Control Circuit/Open
P2661	"B" Rocker Arm Actuator System Performance or Stuck Off
P2662	"B" Rocker Arm Actuator System Stuck On
P2663	"B" Rocker Arm Actuator Control Circuit Low
P2664	"B" Rocker Arm Actuator Control Circuit High
P2665	Fuel Shutoff Valve "B" Control Circuit/Open
P2666	Fuel Shutoff Valve "B" Control Circuit Low
P2667	Fuel Shutoff Valve "B" Control Circuit High
P2668	Fuel Mode Indicator Lamp Control Circuit
P2669	Actuator Supply Voltage "B" Circuit /Open
P2670	Actuator Supply Voltage "B" Circuit Low
P2671	Actuator Supply Voltage "B" Circuit High
P2700	Transmission Friction Element "A" Apply Time Range/Performance
P2701	Transmission Friction Element "B" Apply Time Range/Performance
P2702	Transmission Friction Element "C" Apply Time Range/Performance

Vehicle CODE LOG

DATE	
Vehicle Info	
Vehicle CODE	P:
Current problems	
Solution	
NOTE	

P2703	Transmission Friction Element "D" Apply Time Range/Performance
P2704	Transmission Friction Element "E" Apply Time Range/Performance
P2705	Transmission Friction Element "F" Apply Time Range/Performance
P2706	Shift Solenoid "F"
P2707	Shift Solenoid "F" Performance or Stuck Off
P2708	Shift Solenoid "F" Stuck On
P2709	Shift Solenoid "F" Electrical
P2710	Shift Solenoid "F" Intermittent
P2711	Unexpected Mechanical Gear Disengagement
P2712	Hydraulic Power Unit Leakage
P2713	Pressure Control Solenoid "D"
P2714	Pressure Control Solenoid "D" Performance or Stuck Off
P2715	Pressure Control Solenoid "D" Stuck On
P2716	Pressure Control Solenoid "D" Electrical
P2717	Pressure Control Solenoid "D" Intermittent
P2718	Pressure Control Solenoid "D" Control Circuit / Open
P2719	Pressure Control Solenoid "D" Control Circuit Range/Performance
P2720	Pressure Control Solenoid "D" Control Circuit Low
P2721	Pressure Control Solenoid "D" Control Circuit High
P2722	Pressure Control Solenoid "E"
P2723	Pressure Control Solenoid "E" Performance or Stuck Off
P2724	Pressure Control Solenoid "E" Stuck On
P2725	Pressure Control Solenoid "E" Electrical
P2726	Pressure Control Solenoid "E" Intermittent
P2727	Pressure Control Solenoid "E" Control Circuit / Open
P2728	Pressure Control Solenoid "E" Control Circuit Range/Performance
P2729	Pressure Control Solenoid "E" Control Circuit Low
P2730	Pressure Control Solenoid "E" Control Circuit High
P2731	Pressure Control Solenoid "F"
P2732	Pressure Control Solenoid "F" Performance or Stuck Off
P2733	Pressure Control Solenoid "F" Stuck On
P2734	Pressure Control Solenoid "F" Electrical
P2735	Pressure Control Solenoid "F" Intermittent

Vehicle CODE LOG

DATE	
Vehicle Info	
Vehicle CODE	P:
Current problems	
Solution	
NOTE	

P2736	Pressure Control Solenoid "F" Control Circuit/Open
P2737	Pressure Control Solenoid "F" Control Circuit Range/Performance
P2738	Pressure Control Solenoid "F" Control Circuit Low
P2739	Pressure Control Solenoid "F" Control Circuit High
P2740	Transmission Fluid Temperature Sensor "B" Circuit"
P2741	Transmission Fluid Temperature Sensor "B" Circuit Range Performance
P2742	Transmission Fluid Temperature Sensor "B" Circuit Low
P2743	Transmission Fluid Temperature Sensor "B" Circuit High
P2744	Transmission Fluid Temperature Sensor "B" Circuit Intermittent
P2745	Intermediate Shaft Speed Sensor "B" Circuit
P2746	Intermediate Shaft Speed Sensor "B" Circuit Range/Performance
P2747	Intermediate Shaft Speed Sensor "B" Circuit No Signal
P2748	Intermediate Shaft Speed Sensor "B" Circuit Intermittent
P2749	Intermediate Shaft Speed Sensor "C" Circuit
P2750	Intermediate Shaft Speed Sensor "C" Circuit Range/Performance
P2751	Intermediate Shaft Speed Sensor "C" Circuit No Signal
P2752	Intermediate Shaft Speed Sensor "C" Circuit Intermittent
P2753	Transmission Fluid Cooler Control Circuit/Open
P2754	Transmission Fluid Cooler Control Circuit Low
P2755	Transmission Fluid Cooler Control Circuit High
P2756	Torque Converter Clutch Pressure Control Solenoid
P2757	Torque Converter Clutch Pressure Control Solenoid Control Circuit Performance or Stuck Off
P2758	Torque Converter Clutch Pressure Control Solenoid Control Circuit Stuck On
P2759	Torque Converter Clutch Pressure Control Solenoid Control Circuit Electrical
P2760	Torque Converter Clutch Pressure Control Solenoid Control Circuit Intermittent
P2761	Torque Converter Clutch Pressure Control Solenoid Control Circuit/Open
P2762	Torque Converter Clutch Pressure Control Solenoid Control Circuit Range/Performance
P2763	Torque Converter Clutch Pressure Control Solenoid Control Circuit High
P2764	Torque Converter Clutch Pressure Control Solenoid Control Circuit Low
P2765	Input/Turbine Speed Sensor "B" Circuit
P2766	Input/Turbine Speed Sensor "B" Circuit Range/Performance
P2767	Input/Turbine Speed Sensor "B" Circuit No Signal

Vehicle CODE LOG

DATE	
Vehicle Info	
Vehicle CODE	P:
Current problems	
Solution	
NOTE	

P2768	Input/Turbine Speed Sensor "B" Circuit Intermittent
P2769	Torque Converter Clutch Circuit Low
P2770	Torque Converter Clutch Circuit High
P2771	Four Wheel Drive (4WD) Low Switch Circuit
P2772	Four Wheel Drive (4WD) Low Switch Circuit Range/Performance
P2773	Four Wheel Drive (4WD) Low Switch Circuit Low
P2774	Four Wheel Drive (4WD) Low Switch Circuit High
P2775	Upshift Switch Circuit Range/Performance
P2776	Upshift Switch Circuit Low
P2777	Upshift Switch Circuit High
P2778	Upshift Switch Circuit Intermittent/Erratic
P2779	Downshift Switch Circuit Range/Performance
P2780	Downshift Switch Circuit Low
P2781	Downshift Switch Circuit High
P2782	Downshift Switch Circuit Intermittent/Erratic
P2783	Torque Converter Temperature Too High
P2784	Input/Turbine Speed Sensor "A"/"B" Correlation
P2785	Clutch Actuator Temperature Too High
P2786	Gear Shift Actuator Temperature Too High
P2787	Clutch Temperature Too High
P2788	Auto Shift Manual Adaptive Learning at Limit
P2789	Clutch Adaptive Learning at Limit
P2790	Gate Select Direction Circuit
P2791	Gate Select Direction Circuit Low
P2792	Gate Select Direction Circuit High
P2793	Gear Shift Direction Circuit
P2794	Gear Shift Direction Circuit Low
P2795	Gear Shift Direction Circuit High
P2A00	O2 Sensor Circuit Range/Performance
P2A01	O2 Sensor Circuit Range/Performance
P2A02	O2 Sensor Circuit Range/Performance
P2A03	O2 Sensor Circuit Range/Performance
P2A04	O2 Sensor Circuit Range/Performance

Vehicle CODE LOG

DATE	
Vehicle Info	
Vehicle CODE	P:
Current problems	
Solution	
NOTE	

P2A05	O2 Sensor Circuit Range/Performance
P3400	Cylinder Deactivation System
P3401	Cylinder 1 Deactivation/Intake Valve Control Circuit/Open
P3402	Cylinder 1 Deactivation/Intake Valve Control Performance
P3403	Cylinder 1 Deactivation/Intake Valve Control Circuit Low
P3404	Cylinder 1 Deactivation/Intake Valve Control Circuit High
P3405	Cylinder 1 Exhaust Valve Control Circuit/Open
P3406	Cylinder 1 Exhaust Valve Control Performance
P3407	Cylinder 1 Exhaust Valve Control Circuit Low
P3408	Cylinder 1 Exhaust Valve Control Circuit High
P3409	Cylinder 2 Deactivation/Intake Valve Control Circuit/Open
P3410	Cylinder 2 Deactivation/Intake Valve Control Performance
P3411	Cylinder 2 Deactivation/Intake Valve Control Circuit Low
P3412	Cylinder 2 Deactivation/Intake Valve Control Circuit High
P3413	Cylinder 2 Exhaust Valve Control Circuit/Open
P3414	Cylinder 2 Exhaust Valve Control Performance
P3415	Cylinder 2 Exhaust Valve Control Circuit Low
P3416	Cylinder 2 Exhaust Valve Control Circuit High
P3417	Cylinder 3 Deactivation/Intake Valve Control Circuit/Open
P3418	Cylinder 3 Deactivation/Intake Valve Control Performance
P3419	Cylinder 3 Deactivation/Intake Valve Control Circuit Low
P3420	Cylinder 3 Deactivation/Intake Valve Control Circuit High
P3421	Cylinder 3 Exhaust Valve Control Circuit/Open
P3422	Cylinder 3 Exhaust Valve Control Performance
P3423	Cylinder 3 Exhaust Valve Control Circuit Low
P3424	Cylinder 3 Exhaust Valve Control Circuit High
P3425	Cylinder 4 Deactivation/Intake Valve Control Circuit/Open
P3426	Cylinder 4 Deactivation/Intake Valve Control Performance
P3427	Cylinder 4 Deactivation/Intake Valve Control Circuit Low
P3428	Cylinder 4 Deactivation/Intake Valve Control Circuit High
P3429	Cylinder 4 Exhaust Valve Control Circuit/Open
P3430	Cylinder 4 Exhaust Valve Control Performance
P3431	Cylinder 4 Exhaust Valve Control Circuit Low

Vehicle CODE LOG

DATE	
Vehicle Info	
Vehicle CODE	P:
Current problems	
Solution	
NOTE	

P3432	Cylinder 4 Exhaust Valve Control Circuit High
P3433	Cylinder 5 Deactivation/Intake Valve Control Circuit/Open
P3434	Cylinder 5 Deactivation/Intake Valve Control Performance
P3435	Cylinder 5 Deactivation/Intake Valve Control Circuit Low
P3436	Cylinder 5 Deactivation/Intake Valve Control Circuit High
P3437	Cylinder 5 Exhaust Valve Control Circuit/Open
P3438	Cylinder 5 Exhaust Valve Control Performance
P3439	Cylinder 5 Exhaust Valve Control Circuit Low
P3440	Cylinder 5 Exhaust Valve Control Circuit High
P3441	Cylinder 6 Deactivation/Intake Valve Control Circuit/Open
P3442	Cylinder 6 Deactivation/Intake Valve Control Performance
P3443	Cylinder 6 Deactivation/Intake Valve Control Circuit Low
P3444	Cylinder 6 Deactivation/Intake Valve Control Circuit High
P3445	Cylinder 6 Exhaust Valve Control Circuit/Open
P3446	Cylinder 6 Exhaust Valve Control Performance
P3447	Cylinder 6 Exhaust Valve Control Circuit Low
P3448	Cylinder 6 Exhaust Valve Control Circuit High
P3449	Cylinder 7 Deactivation/Intake Valve Control Circuit/Open
P3450	Cylinder 7 Deactivation/Intake Valve Control Performance
P3451	Cylinder 7 Deactivation/Intake Valve Control Circuit Low
P3452	Cylinder 7 Deactivation/Intake Valve Control Circuit High
P3453	Cylinder 7 Exhaust Valve Control Circuit/Open
P3454	Cylinder 7 Exhaust Valve Control Performance
P3455	Cylinder 7 Exhaust Valve Control Circuit Low
P3456	Cylinder 7 Exhaust Valve Control Circuit High
P3457	Cylinder 8 Deactivation/Intake Valve Control Circuit/Open
P3458	Cylinder 8 Deactivation/Intake Valve Control Performance
P3459	Cylinder 8 Deactivation/Intake Valve Control Circuit Low
P3460	Cylinder 8 Deactivation/Intake Valve Control Circuit High
P3461	Cylinder 8 Exhaust Valve Control Circuit/Open
P3462	Cylinder 8 Exhaust Valve Control Performance
P3463	Cylinder 8 Exhaust Valve Control Circuit Low
P3464	Cylinder 8 Exhaust Valve Control Circuit High

Vehicle CODE LOG

DATE	
Vehicle Info	
Vehicle CODE	P:
Current problems	
Solution	
NOTE	

P3465	Cylinder 9 Deactivation/Intake Valve Control Circuit/Open
P3466	Cylinder 9 Deactivation/Intake Valve Control Performance
P3467	Cylinder 9 Deactivation/Intake Valve Control Circuit Low
P3468	Cylinder 9 Deactivation/Intake Valve Control Circuit High
P3469	Cylinder 9 Exhaust Valve Control Circuit/Open
P3470	Cylinder 9 Exhaust Valve Control Performance
P3471	Cylinder 9 Exhaust Valve Control Circuit Low
P3472	Cylinder 9 Exhaust Valve Control Circuit High
P3473	Cylinder 10 Deactivation/Intake Valve Control Circuit/Open
P3474	Cylinder 10 Deactivation/Intake Valve Control Performance
P3475	Cylinder 10 Deactivation/Intake Valve Control Circuit Low
P3476	Cylinder 10 Deactivation/Intake Valve Control Circuit High
P3477	Cylinder 10 Exhaust Valve Control Circuit/Open
P3478	Cylinder 10 Exhaust Valve Control Performance
P3479	Cylinder 10 Exhaust Valve Control Circuit Low
P3480	Cylinder 10 Exhaust Valve Control Circuit High
P3481	Cylinder 11 Deactivation/Intake Valve Control Circuit/Open
P3482	Cylinder 11 Deactivation/Intake Valve Control Performance
P3483	Cylinder 11 Deactivation/Intake Valve Control Circuit Low
P3484	Cylinder 11 Deactivation/Intake Valve Control Circuit High
P3485	Cylinder 11 Exhaust Valve Control Circuit/Open
P3486	Cylinder 11 Exhaust Valve Control Performance
P3487	Cylinder 11 Exhaust Valve Control Circuit Low
P3488	Cylinder 11 Exhaust Valve Control Circuit High
P3489	Cylinder 12 Deactivation/Intake Valve Control Circuit/Open
P3490	Cylinder 12 Deactivation/Intake Valve Control Performance
P3491	Cylinder 12 Deactivation/Intake Valve Control Circuit Low
P3492	Cylinder 12 Deactivation/Intake Valve Control Circuit High
P3493	Cylinder 12 Exhaust Valve Control Circuit/Open
P3494	Cylinder 12 Exhaust Valve Control Performance
P3495	Cylinder 12 Exhaust Valve Control Circuit Low
P3496	Cylinder 12 Exhaust Valve Control Circuit High
P3497	Cylinder Deactivation System

Vehicle CODE LOG

DATE	
Vehicle Info	
Vehicle CODE	P:
Current problems	
Solution	
NOTE	

www.ingramcontent.com/pod-product-compliance
Lightning Source LLC
Chambersburg PA
CBHW080503220526
45465CB00006B/2357